Lecture Notes in Computer Science 9940

Commenced Publication in 1973
Founding and Former Series Editors:
Gerhard Goos, Juris Hartmanis, and Jan van Leeuwen

More information about this series at http://www.springer.com/series/8637

Abdelkader Hameurlain · Josef Küng
Roland Wagner · Qimin Chen (Eds.)

Transactions on Large-Scale Data- and Knowledge-Centered Systems XXVIII

Special Issue on Database- and Expert-Systems Applications

Springer

Editors-in-Chief

Abdelkader Hameurlain
IRIT, Paul Sabatier University
Toulouse
France

Roland Wagner
FAW, University of Linz
Linz
Austria

Josef Küng
FAW, University of Linz
Linz
Austria

Guest Editor

Qimin Chen
HP Labs
Sunnyvale, CA
USA

ISSN 0302-9743   ISSN 1611-3349 (electronic)
Lecture Notes in Computer Science
ISBN 978-3-662-53454-0   ISBN 978-3-662-53455-7 (eBook)
DOI 10.1007/978-3-662-53455-7

Library of Congress Control Number: 2015943846

This Springer imprint is published by Springer Nature
The registered company is Springer-Verlag GmbH Berlin Heidelberg

Preface

The 26th International Conference on Database and Expert Systems Applications, DEXA 2015, held in Valencia, Spain, September 1–4, 2015, provided a premier forum and unique opportunity for researchers, developers, and users from different disciplines to present the state of the art, exchange research ideas, share industry experiences, and explore future directions at the intersection of data management, knowledge engineering, and artificial intelligence. This special issue of Springer's *Transactions on Large-Scale Data- and Knowledge-Centered Systems (TLDKS)* contains extended versions of selected papers presented at the conference. While these articles describe the technical trend and the breakthroughs made in the field, the general message delivered from them is that turning big data to big value requires incorporating cutting-edge hardware, software, algorithms and machine-intelligence.

Efficient graph-processing is a pressing demand in social-network analytics. A solution to the challenge of leveraging modern hardware in order to speed up the similarity join in graph processing is given in the article "Accelerating Set Similarity Joins Using GPUs", authored by Mateus S. H. Cruz, Yusuke Kozawa, Toshiyuki Amagasa, and Hiroyuki Kitagawa. In this paper, the authors propose a GPU (Graphics Processing Unit) supported set similarity joins scheme. It takes advantage of the massive parallel processing offered by GPUs, as well as the space efficiency of the MinHash algorithm in estimating set similarity, to achieve high performance without sacrificing accuracy. The experimental results show more than two orders of magnitude performance gain compared with the serial version of CPU implementation, and 25 times performance gain compared with the parallel version of CPU implementation. This solution can be applied to a variety of applications such as data integration and plagiarism detection.

Parallel processing is the key to accelerating machine-learning on big data. However, many machine leaning algorithms involve iterations that are hard to be parallelized from either the load balancing among processors, memory access overhead, or race conditions, such as those relying on hierarchical parameter estimation. The article "Divide-and-Conquer Parallelism for Learning Mixture Models", authored by Takaya Kawakatsu, Akira Kinoshita, Atsuhiro Takasu, and Jun Adachi, addresses this problem. In this paper, the authors propose a recursive divide-and-conquer-based parallelization method for high-speed machine learning, which uses a tree structure for recursive tasks to enable effective load balancing and to avoid race conditions in memory access. The experiment results show that applying this mechanism to machine learning can reach a scalability superior to FIFO scheduling, with robust load imbalance.

Maintaining multistore systems has become a new trend for integrated access to multiple, heterogeneous data, either structured or unstructured. A typical solution is to extend a relational query engine to use SQL-like queries to retrieve data from other data sources such as HDFS, which, however, requires the system to provide a relational view of the unstructured data. An alternative approach is proposed in the article "Multistore Big Data Integration with CloudMdsQL", authored by Carlyna

Bondiombouy, Boyan Kolev, Oleksandra Levchenko, and Patrick Valduriez. In this paper, a functional SQL-like query language (based on CloudMdsQL) is introduced for integrated data retrieved from different data stores, therefore taking full advantage of the functionality of the underlying data management frameworks. It allows user defined map/filter/reduce operators to be embedded in traditional SQL statements. It further allows the filtering conditions to be pushed down to the underlying data processing framework as early as possible for the purpose of optimization. The usability of this query language and the benefits of the query optimization mechanism are demonstrated by the experimental results.

One of the primary goals of exploring big data is to discover useful patterns and concepts. There exist several kinds of conventional pattern matching algorithms; for instance, the terminology-based algorithms are used to compare concepts based on their names or descriptions, the structure-based algorithms are used to align concept hierarchies to find similarities; the statistic-based algorithms classify concepts in terms of various generative models. In the article "Ontology Matching with Knowledge Rules", authored by Shangpu Jiang, Daniel Lowd, Sabin Kafle, and Dejing Dou, the focus is shifted to aligning concepts by comparing their relationships with other known concepts. Such relationships are expressed in various ways – Bayesian networks, decision trees, association rules, etc.

The article "Regularized Cost-Model Oblivious Database Tuning with Reinforcement Learning", authored by Debabrota Basu, Qian Lin, Weidong Chen, Hoang Tam Vo, Zihong Yuan, Pierre Senellart, and Stephane Bressan, proposes a machine learning approach for adaptive database performance tuning, a critical issue for efficient information management, especially in the big data context. With this approach, the cost model is learned through reinforcement learning. In the use case of index tuning, the executions of queries and updates are modeled as a Markov decision process, with states represented in database configurations, actions causing configuration changes, corresponding cost parameters, as well as query and update evaluations. Two important challenges in the reinforcement learning process are discussed: the unavailability of a cost model and the size of the state space. The solution to the first challenge is to learn the cost model iteratively, using regularization to avoid overfitting; the solution to the second challenge is to prune the state space intelligently. The proposed approach is empirically and comparatively evaluated on a standard OLTP dataset, which shows competitive advantage.

The article "Workload-Aware Self-tuning Histograms for the Semantic Web", authored by Katerina Zamani, Angelos Charalambidis, Stasinos Konstantopoulos, Nickolas Zoulis, and Effrosyni Mavroudi, further discusses how to optimize the histograms for semantic Web. As we know, query processing systems typically rely on histograms which represent approximate data distribution, to optimize query execution. Histograms can be constructed by scanning the datasets and aggregating the values of the selected fields, and progressively refined by analyzing query results. This article tackles the following issue: histograms are typically built from numerical data, but the Semantic Web is described with various data types which are not necessarily numeric. In this work a generalized histograms framework over arbitrary data types is established with the formalism for specifying value ranges corresponding to various datatypes. Then the Jaro-Winkler metric is introduced to define URI ranges based on the hierarchical nature of URI strings. The empirical evaluation results, conducted using the open-sourced STRHist system that implements this approach, demonstrate its competitive advantage.

We would like to thank all the authors for their contributions to this special issue. We are grateful to the reviewers of these articles for their invaluable efforts in collaborating with the authors to deliver readers the precise ideas, theories, and solutions on the above state-of-the-art technologies. Our deep appreciation also goes to Prof. Roland Wagner, Chairman of the DEXA Organization, Ms. Gabriela Wagner, Secretary of DEXA, the distinguished keynote speakers, Program Committee members, and all presenters and attendees of DEXA 2015. Their contributions help to keep DEXA a distinguished platform for exchanging research ideas and exploring new directions, thus setting the stage for this special TLDKS issue.

June 2016   Qiming Chen
                                                           Abdelkader Hameurlain

Organization

Editorial Board

Contents

Contents

Accelerating Set Similarity Joins Using GPUs

Mateus S.H. Cruz[1]([✉]), Yusuke Kozawa[1], Toshiyuki Amagasa[2],
and Hiroyuki Kitagawa[2]

[1] Graduate School of Systems and Information Engineering,
University of Tsukuba, Tsukuba, Japan
{mshcruz,kyusuke}@kde.cs.tsukuba.ac.jp
[2] Faculty of Engineering, Information and Systems,
University of Tsukuba, Tsukuba, Japan
{amagasa,kitagawa}@cs.tsukuba.ac.jp

Abstract. We propose a scheme for efficient set similarity joins on
Graphics Processing Units (GPUs). Due to the rapid growth and diversification of data, there is an increasing demand for fast execution of
set similarity joins in applications that vary from data integration to
plagiarism detection. To tackle this problem, our solution takes advantage of the massive parallel processing offered by GPUs. Additionally,
we employ MinHash to estimate the similarity between two sets in terms
of Jaccard similarity. By exploiting the high parallelism of GPUs and
the space efficiency provided by MinHash, we can achieve high performance without sacrificing accuracy. Experimental results show that our
proposed method is more than two orders of magnitude faster than the
serial version of CPU implementation, and 25 times faster than the parallel version of CPU implementation, while generating highly precise query
results.

Keywords: GPU · Parallel processing · Similarity join · MinHash

1 Introduction

A *similarity join* is an operator that, given two database relations and a similarity threshold, outputs all pairs of records, one from each relation, whose similarity is greater than the specified threshold. It has become a significant class
of database operations due to the diversification of data, and it is used in many
applications, such as data cleaning, entity recognition and duplicate elimination [3,5]. As an example, for data integration purposes, it might be interesting
to detect whether *University of Tsukuba* and *Tsukuba University* refer to the
same entity. In this case, the similarity join can identify such a pair of records
as being similar.

Set similarity join [11] is a variation of similarity join that works on sets
instead of regular records, and it is an important operation in the family of
similarity joins due to its applicability on different data (e.g., market basket
data, text and images). Regarding the similarity aspect, there is a number of

© Springer-Verlag Berlin Heidelberg 2016
A. Hameurlain et al. (Eds.): TLDKS XXVIII, LNCS 9940, pp. 1–22, 2016.
DOI: 10.1007/978-3-662-53455-7_1

well-known similarity metrics used to compare sets (e.g., Jaccard similarity and cosine similarity).

One of the major drawbacks of a set similarity join is that it is a computationally demanding task, especially in the current scenario in which the size of datasets grows rapidly due to the trend of Big Data. For this reason, many researchers have proposed different set similarity join processing schemes [21,23,24]. Among them, it has been shown that parallel computation is a cost-effective option to tackle this problem [16,20], especially with the use of Graphics Processing Units (GPUs), which have been gaining much attention due to their performance in general processing [19].

There are numerous technical challenges when performing set similarity join using GPUs. First, how to deal with large datasets using GPU's memory, which is limited up to a few GBs in size. Second, how to make the best use of the high parallelism of GPUs in different stages of the processing (e.g., similarity computation and the join itself). Third, how to take advantage of the different types of memories on GPUs, such as device memory and shared memory, in order to maximize the performance.

In this research, we propose a new scheme of set similarity join on GPUs. To address the aforementioned technical challenges, we employ MinHash [2] to estimate the similarity between two sets in terms of their Jaccard similarity. MinHash is known to be a space-efficient algorithm to estimate the Jaccard similarity, while making it possible to maintain a good trade-off between accuracy and computation time. Moreover, we carefully design data structures and memory access patterns to exploit the GPU's massive parallelism and achieve high speedups.

Experimental results show that our proposed method is more than two orders of magnitude faster than the serial version of CPU implementation, and 25 times faster than the parallel version of CPU implementation. In both cases, we assure the quality of the results by maximizing precision and recall values. We expect that such contributions can be effectively applied to process large datasets in real-world applications.

This paper extends a previous work [25] by exploring the state of the art in more depth, by providing more details related to implementation and methodology, and by offering additional experiments.

The remainder of this paper is organized as follows. Section 2 offers an overview of the similarity join operation applied to sets. Section 3 introduces the special hardware used, namely GPU, highlighting its main features and justifying its use in this work. In Sect. 4, we discuss the details of the proposed solution, and in Sect. 5 we present the experiments conducted to evaluate it. Section 6 examines the related work. Finally, Sect. 7 covers the conclusions and the future work.

2   Similarity Joins over Sets

In a database, given two relations containing many records, it is common to use the join operation to identify the pairs of records that are similar enough to satisfy a predefined similarity condition. Such operation is called similarity join. This section introduces the application of similarity joins over sets, as well as the similarity measure used in our work, namely Jaccard similarity. After that, we explain how we take advantage of the MinHash [2] technique to estimate similarities, thus saving space and reducing computation time.

2.1 Set Similarity Joins

In many applications, we need to deal with sets (or multisets) of values as a part of data records. Some of the major examples are bag-of-words (documents), bag-of-visual-words (images) and transaction data [1,15]. Given database relations with records containing sets, one may wish to identify pairs of records whose sets are similar; in other words, two sets that share many elements. We refer to this variant of similarity join as a *set similarity join*. Henceforth, we use similarity join to denote set similarity join, if there is no ambiguity.

For example, Fig. 1 presents two collections of documents (R and S) that contain two documents each (R_0, R_1; S_0, S_1). In this scenario, the objective of the similarity join is to retrieve pairs of documents, one from each relation, that have a similarity degree greater than a specified threshold. Although there is a variety of methods to calculate the similarity between two documents, here we represent documents as sets of words (or *tokens*), and apply a set similarity method to determine how similar they are. We choose to use the Jaccard similarity (JS) since it is a well-known and commonly used technique to measure similarity between sets, and its calculation has high affinity with the GPU architecture. One can calculate the Jaccard similarity between two sets, X and Y, in the following way: $JS(X, Y) = |X \cap Y|/|X \cup Y|$. Considering this formula and the documents in Fig. 1, we obtain the following results: $JS(R_0, S_0) = 3/5 = 0.6$, $JS(R_0, S_1) = 1/6 = 0.17$, $JS(R_1, S_0) = 1/7 = 0.14$ and $JS(R_1, S_1) = 1/6 = 0.17$.

The computation of Jaccard similarity requires a number of pairwise comparisons among the elements from different sets to identify common elements, which incurs a long execution time, particularly when the sets being compared are large. In addition, it is necessary to store the whole sets in memory, which can require prohibitive storage [13].

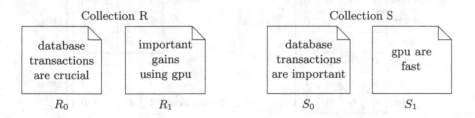

Fig. 1. Two collections of documents (R and S).

2.2   MinHash

To address the aforementioned problems, Broder et al. proposed a technique called MinHash (Min-wise Hashing) [2]. Its main idea is to create signatures for each set based on its elements and then compare the signatures to estimate their Jaccard similarity. If two sets have many coinciding signature parts, they share some degree of similarity. In this way, it is possible to estimate the Jaccard similarity without conducting costly scans over all elements. In addition one only needs to store the signatures instead of all the elements of the sets, which greatly contributes to reduce storage space.

After its introduction, Li et al. suggested a series of improvements for the MinHash technique related to memory use and computation performance [12–14]. Our work is based on the latest of those improvements, namely, One Permutation Hashing [14].

In order to estimate the similarity of the documents in Fig. 1 using One Permutation Hashing, first we change their representation to a data structure called *characteristic matrix* (Fig. 2a), which assigns the value *1* when a token represented by a row belongs to a document represented by a column, and *0* when it does not.

After that, in order to obtain an unbiased similarity estimation, a random permutation of rows is applied to the characteristic matrix, followed by a division of the rows into partitions (henceforth called *bins*) of approximate size (Fig. 2b). However, the actual permutation of rows in a large matrix constitutes an expensive operation, and MinHash uses hash functions to emulate such permutation. Compared to the original MinHash approach [2], One Permutation Hashing presents a more efficient strategy for computation and storage, since it computes only one permutation instead of a few hundreds. For example, considering a dataset with D (e.g., 10^9) features, each permutation emulated by

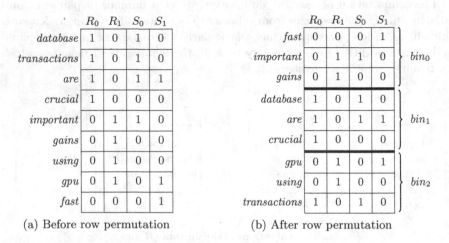

(a) Before row permutation   (b) After row permutation

Fig. 2. Characteristic matrices constructed based on the documents from Fig. 1, before and after a permutation of rows.

	b_0	b_1	b_2
R_0	*	3	8
R_1	1	*	6
S_0	1	3	8
S_1	0	4	6

Fig. 3. Signature matrix, with columns corresponding to the bins composing the signatures of documents, and rows corresponding to the documents themselves. The symbol * denotes an empty bin.

a hash function would require a an array of D positions. Considering a large number k (e.g., $k = 500$) of hash functions, a total of $D \times k$ positions would be needed for the scheme, thus making the storage requirements impractical for many large-scale applications [14].

For each bin, each document has a value that will compose its signature. This value is the index of the row containing the first 1 (scanning the matrix in a top-down fashion) in the column representing the document. For example, the signature for the document S_0 is 1, 3 and 8. It can happen that a bin for a given document does not have any value (e.g., the first bin of set R_0, since it has no 1), and this case is also taken into consideration during the similarity estimation. Figure 3 shows a data structure called *signature matrix*, which contains the signatures obtained for all the documents.

Finally, the similarity between any two documents is estimated by Eq. 1 [14], where N_{mat} is the number of matching bins between the signatures of the two documents, b represents the total number of bins composing the signatures, and N_{emp} refers to the number of matching empty bins.

$$Sim(X, Y) = \frac{N_{mat}}{(b - N_{emp})} \tag{1}$$

The estimated similarities for the given example are $Sim(R_0, S_0) = 2/3 = 0.6$, $Sim(R_0, S_1) = 0/3 = 0$, $Sim(R_1, S_0) = 1/3 = 0.33$ and $Sim(R_1, S_1) = 1/3 = 0.33$. Even though this is a simple example, the estimated values can be considered close to the real Jaccard similarities previously calculated (0.67, 0.17, 0.14 and 0.17). In practical terms, using more bins yields a more accurate estimation, but it also increases the size of the signature matrix.

Let us observe an important characteristic of MinHash. Since the signatures are independent of each other, it presents a good opportunity for parallelization. Indeed, the combination of MinHash and parallel processing using GPUs has been considered by Li et al. [13], as they showed a reduction of the processing time by more than an order of magnitude in online learning applications. While their focus was the MinHash itself, here we use it as a tool in the similarity join processing.

3   General-Purpose Processing on Graphics Processing Units

Despite being originally designed for games and other graphic applications, the applications of Graphics Processing Units (GPUs) have been extended to general computation due to their high computational power [19]. This section presents the features of this hardware and the challenges encountered when using it.

The properties of a modern GPU can be seen from both a computing and a memory-related perspective (Fig. 4). In terms of computational components, the GPU's *scalar processors* (SPs) run the primary processing unit, called *thread*. GPU programs (commonly referred to as *kernels*) run in an SPMD (Single Program Multiple Data) fashion on these lightweight threads. Threads form *blocks*, which are scheduled to run on *streaming multiprocessors* (SMs).

The memory hierarchy of a GPU consists of three main elements: *registers*, *shared memory* and *device memory*. Each thread has access to its own registers (quickly accessible, but small in size) through the register file, but cannot access the registers of other threads. In order to share data among threads in a block, it is possible to use the shared memory, which is also fast, but still small (16 KB to 96 KB per SM depending on the GPU's capability). Lastly, in order to share data between multiple blocks, the device memory (also called *global memory*) is used. However, it should be noted that the device memory suffers from a long access latency as it resides outside the SMs.

When programming a GPU, one of the greatest challenges is the effective utilization of this hardware's architecture. For example, there are several benefits in exploring the faster memories, as it minimizes the access to the slower device memory and increases the overall performance.

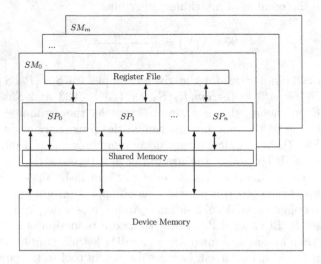

Fig. 4. Architecture of a modern GPU.

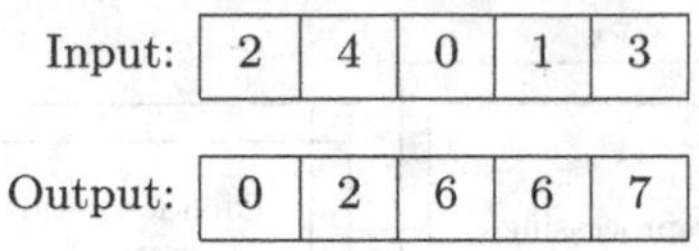

Input: | 2 | 4 | 0 | 1 | 3 |

Output: | 0 | 2 | 6 | 6 | 7 |

Fig. 5. Scan primitive.

In order to apply a GPU for general processing, it is common to use dedicated libraries that can facilitate such task. Our solution employs NVIDIA's CUDA [17], which provides an extension of the C programming language, by which one can define parts of a program to be executed on the GPU.

In terms of algorithms, a number of data-parallel operations, usually called *primitives*, have been ported to be executed on GPUs in order to facilitate programming tasks. He et al. [7,8] provide details on the design and implementation of many of these primitives.

One primitive particularly useful for our work is *scan* or *prefix-sum* (Definition 1 [26]), which has been target of several works [22,27,28]. Figure 5 illustrates its basic form (where the binary operator is addition) by receiving as input an array of integers and outputting an array where the value in each position is the sum of the values of previous positions.

Definition 1. *The* scan *(or* prefix-sum*) operation takes a binary associative operator $\oplus$ with identity* I, *and an array of* n *elements* $[a_0, a_1, ..., a_{n-1}]$, *and returns the array* $[I, a_0, (a_0 \oplus a_1), ..., (a_0 \oplus a_1 \oplus ... \oplus a_{n-2})]$.

As detailed in Sect. 4.3, we use the scan primitive to calculate the positions where each GPU block will write the result of its computation, allowing us to overcome the lack of incremental memory allocation during the execution of kernels and to avoid write conflicts between blocks. We chose to adopt the scan implementation provided by the library Thrust [9] due to its high performance and ease of use.

4 GPU Acceleration of Set Similarity Joins

In the following discussion, we consider sets to be text documents stored on disk, but the solution can be readily adapted to other types of data, as shown in the experimental evaluation (Sect. 5). We also assume that techniques to prepare text data for processing (e.g., stop-word removal and stemming) are out of our scope, and should take place before the similarity join processing.

Figure 6 shows the workflow of the proposed scheme. First, the system receives two collections of documents representing relations R and S. After that, it executes the three main steps of our solution: preprocessing, signature matrix computation and similarity join. Finally, the result can be presented to the user after being properly formatted.

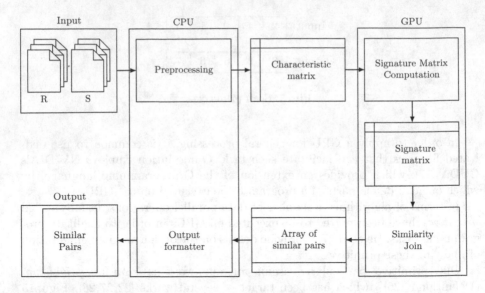

Fig. 6. System's workflow.

4.1   Preprocessing

In the preprocessing step, we construct a compact representation of the characteristic matrix, since the original one is usually highly sparse. By doing so, the data to be transferred to the GPU is greatly reduced (more than 95 % for the datasets used in the experimental evaluation in Sect. 5).

This representation is based on the Compressed Row Storage (CRS) format [6], which uses three arrays: *var*, which stores the values of the nonzero elements of the matrix; *col_ind*, that holds the column indexes of the elements in the *var* array; and *row_ptr*, which keeps the locations in the *var* array that start a row in the matrix.

Considering that the nonzero elements of the characteristic matrix have the same value, *1*, there is only need to store their positions. Figure 7 shows such representation for the characteristic matrix of the previous example (Fig. 2). The array *doc_start* holds the positions in the array *doc_tok* where the documents start, and the array *doc_tok* shows what tokens belong to each document.

	R_0	R_1	S_0	S_1										
doc_start	0	4	8	12	15									
doc_tok	0	1	2	3	4	5	6	7	0	1	2	4	2	7	8

Fig. 7. Compact representation of the characteristic matrix.

After its construction, the characteristic matrix is sent to the GPU, and we assume it fits completely in the device memory. The processing of large datasets, which do not fit into the device memory is part of future work. Nevertheless, the aforementioned method allows us to deal with sufficiently large datasets using current GPUs in many practical applications.

4.2 Signature Matrix Computation on GPU

Once the characteristic matrix is in the GPU's device memory, the next step is to construct the signature matrix. Algorithm 1 shows how we parallelize the MinHash technique, and Fig. 8 illustrates such processing. In practical terms, one block is responsible for computing the signature of one document at a time. Each thread in the block (1) accesses the device memory, (2) retrieves the position of one token of the document, (3) applies a hash function to it to simulate the row permutation, (4) calculates which bin the token will fit into, and (5) updates that bin. If more than one value is assigned to the same bin, the algorithm keeps the minimum value (hence the name MinHash).

During its computation, the signature for the document is stored in the shared memory, which supports fast communication between the threads of a block. This is advantageous in two aspects: (1) it allows fast updates of values when constructing the signature matrix, and (2) since different threads can

---

Algorithm 1. Parallel MinHash.

---

    input : characteristic matrix $CM_{t \times d}$ (t tokens, d documents), number of bins b
    output: signature matrix $SM_{d \times b}$ (d documents, b bins)
1 bin_size $\leftarrow \lfloor t/b \rfloor$;
2 **for** $i \leftarrow 0$ *to* d **in parallel do** // executed by blocks
3   **for** $j \leftarrow 0$ *to* t **in parallel do** // executed by threads
4   **if** $CM_{j,i} = 1$ **then**
5             $h \leftarrow hash(CM_{j,i})$;
6             $bin\_idx \leftarrow \lfloor h/bin\_size \rfloor$;
7             $SM_{i,bin\_idx} \leftarrow min(SM_{i,bin\_idx}, h)$;
8   **end**
9   **end**
10 **end**

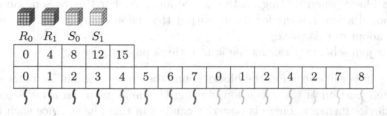

R_0	R_1	S_0	S_1										
0	4	8	12	15									
0	1	2	3	4	5	6	7	0	1	2	4	2	7	8

Fig. 8. Computation of the signature matrix based on the characteristic matrix. Each GPU block is responsible for one document, and each thread is assigned to one token.

access sequential memory positions, it favors the coalesced access to the device memory when the signature computation ends. Accessing the device memory in a coalesced manner means that a number of threads will access consecutive memory locations, and such accesses can be grouped into a single transaction. This makes the transfer of data from and to the device memory significantly faster.

The complete signature matrix is laid out in the device memory as a single array of integers. Since the number of bins per signature is known, it is possible to perform direct access to the signature of any given document.

After the signature matrix is constructed, it is kept in the GPU's memory to be used in the next step: the join itself. This also minimizes data transfers between CPU and GPU.

4.3   Similarity Joins on GPU

The next step is the similarity join, and it utilizes the results obtained in the previous phase, i.e., the signatures generated using MinHash. To address the similarity join problem, we choose to parallelize the nested-loop join (NLJ) algorithm. The nested-loop join algorithm iterates through the two relations being joined and check whether the pairs of records, one from each relation, comply with a given predicate. For the similarity join case, this predicate is that the records of the pairs must have a degree of similarity greater than a given threshold.

Algorithm 2 outlines our parallelization of the NLJ for GPUs. Initially, each block reads the signature of a document from collection R and copies it to the shared memory (line 2, Fig. 9a). Then, threads compare the value of each bin of that signature to the corresponding signature bin of a document from collection S (lines 3–7), checking whether they match and whether the bin is empty (lines 8–12). The access to the data in the device memory is done in a coalesced manner, as illustrated by Fig. 9b. Finally, using Eq. 1, if the comparison yields a similarity greater than the given threshold (line 15–16), that pair of documents belongs to the final result (line 17).

As highlighted by He et al. [8], outputting the result from a join performed in the GPU raises two main problems. Firstly, since the size of the output is initially unknown, it is also not possible to know how much memory should be allocated on the GPU to hold the result. In addition, there may be conflicts between blocks when writing on the device memory. For this reason, He et al. [8] proposed a join scheme for result output that allows parallel writing, which we also adopt in this work.

Their join scheme performs the join in three phases (Fig. 10):

1. The join is run once, and the blocks count the number of similar pairs found in their portion of the execution, writing this amount in an array stored in the device memory. There is no write conflict in this phase, since each block writes in a different position of the array.

Algorithm 2. Parallel nested-loop join.

---

input : signature matrix $SM_{d \times b}$ (d documents, b bins), similarity threshold ε
output: pairs of sets whose similarity is greater than ε

1 **foreach** $r \in R$ **in parallel do** // executed by blocks
2      $r\_signature \leftarrow SM_r$;// read the row corresponding to the signature
        of r and store it in the shared memory
3   **foreach** $s \in S$ **in parallel do** // executed by threads
4          $coinciding\_minhashes \leftarrow 0$;
5          $empty\_bins \leftarrow 0$;
6   **for** $i \leftarrow 0$ *to* b **do**
7   **if** $r\_signature_i = SM_{s,i}$ **then**
8   **if** $r\_signature_i$ *is empty* **then**
9                      $empty\_bins \leftarrow empty\_bins + 1$;
10   **else**
11                     $coinciding\_minhashes \leftarrow coinciding\_minhashes + 1$;
12   **end**
13   **end**
14   **end**
15         $pair\_similarity \leftarrow coinciding\_minhashes/(b - empty\_bins)$;
16   **if** $pair\_similarity \geq \varepsilon$ **then**
17             $output(r, s)$;
18   **end**
19   **end**
20 **end**

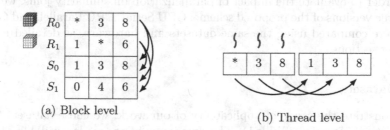

(a) Block level

(b) Thread level

Fig. 9. Parallelization of NLJ.

2. Using the scan primitive, it is possible to know the correct size of memory that should be allocated for the results, as well as where the threads of each block should start writing the similar pairs they found.
3. The similarity join is run once again, outputting the similar pairs to the proper positions in the allocated space.

After that, depending on the application, the pairs can be transferred back to the CPU and output to the user (using the output formatter) or kept in the GPU for further processing by other algorithms.

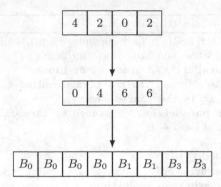

Fig. 10. Example of the three-phase join scheme [8]. First, four blocks write the size of their results in the first array. Then, the scan primitive gives the starting positions where each block should write. Finally, each block writes its results in the last array.

5   Experiments

In this section we present the experiments performed to evaluate our proposal. First, we introduce the used datasets and the environment on which the experiments were conducted. Then we show the results related to performance and accuracy. For all the experiments, unless stated, the similarity threshold was 0.8 and the number of bins composing the sets' signatures was 32.

In order to evaluate the impact of parallelization on similarity joins, we created three versions of the proposed scheme: CPU Serial, CPU Parallel, and GPU. They were compared using the same datasets and hardware, as detailed in the following sections.

5.1   Datasets

To demonstrate the range of applicability of our work, we chose datasets from three distinct domains (Table 1). The *Images* dataset, made available at the UCI Machine Learning Repository[1], consists of image features extracted from the Corel image collection. The *Abstracts* dataset, composed by abstracts of publications from MEDLINE, were obtained from TREC-9 Filtering Track Collections[2]. Finally, *Transactions* is a transactional dataset available through the FIMI repository[3].

From the original datasets, we chose sets uniformly at random in order to create the collections R and S, whose sizes vary from 1,024 to 524,288 sets.

---

[1] http://archive.ics.uci.edu/ml/datasets/.

[2] http://trec.nist.gov/data/t9_filtering.html.

[3] http://fimi.ua.ac.be/data/.

Table 1. Characteristics of datasets.

Dataset	Cardinality	Avg. # of tokens per record
Images	68,040	32
Abstracts	233,445	165
Transactions	1,692,082	177

5.2 Environment

The CPU used in our experiments was an Intel Xeon E5-1650 (6 cores, 12 threads) with 32 GB of memory. The GPU was an NVIDIA Tesla K20Xm (2688 scalar processors) with 6 GB of memory. Regarding the compilers, GCC 4.4.7 (with the flag -O3) was used for the part of the code to run on the CPU, and NVCC 6.5 (with the flags -O3 and -use_fast_math) compiled the code for the GPU. For the parallelization of the CPU version, we used OpenMP 4.0 [18]. The implementation of the hash function was done using MurmurHash [10].

5.3 Performance Comparison

Figures 11, 12 and 13 present the execution time of our approach for the three implementations (GPU, CPU Parallel and CPU Serial) using the three datasets.

Let us first consider the MinHash part, i.e., the time taken for the construction of the signature matrix. It can be seen from the results (Fig. 11a, b and c) that the GPU version of MinHash is more than 20 times faster than the serial implementation on CPU, and more than 3 times faster than the parallel implementation on CPU. These findings reinforce the idea that MinHash is indeed suitable for parallel processing.

For the join part (Fig. 12a, b and c), the speedups are even higher. The GPU implementation is more than 150 times faster than the CPU Serial implementation, and almost 25 times faster than the CPU Parallel implementation.

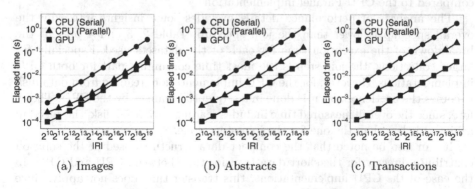

(a) Images   (b) Abstracts   (c) Transactions

Fig. 11. Minhash performance comparison ($|R| = |S|$).

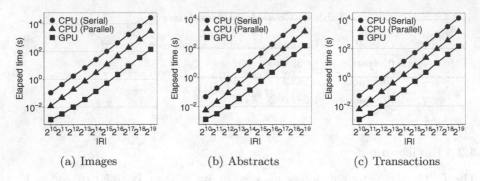

(a) Images   (b) Abstracts   (c) Transactions

Fig. 12. Join performance comparison ($|R| = |S|$).

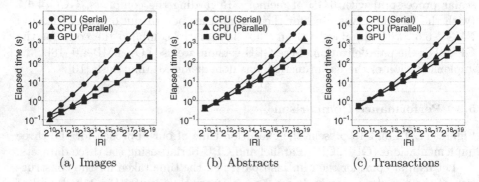

(a) Images   (b) Abstracts   (c) Transactions

Fig. 13. Overall performance comparison ($|R| = |S|$).

The speedups of more than two orders of magnitude demonstrate that the NLJ algorithm can benefit from the massive parallelism provided by GPUs.

Measurements of the total time of execution (Fig. 13a, b and c) show that the GPU implementation achieves speedups of approximately 120 times when compared to the CPU Serial implementation, and approximately 20 times when compared to the CPU Parallel implementation.

The analysis of performance details provides some insights into why the overall speedup is lower than the join speedup. Tables 2, 3 and 4 present the breakdown of the execution time for each of the datasets used. Especially for larger collections, the join step is the most time consuming part for both CPU implementations. However, for the GPU implementation, reading from data disk becomes the bottleneck, as it is done in a sequential manner by the CPU. Therefore, since the overall measured time includes reading data from disk, the speedup achieved is less than the one for the join step alone.

It can also be noted that the compact data structures used in the solution contribute directly for the short data transfer time between CPU and GPU. In the case of the CPU implementations, this transfer time does not apply, since the data stays on the CPU throughout the whole execution.

Table 2. Breakdown of the execution time in seconds when joining collections of the same size (Images dataset, $|R| = |S| = 524,288$).

	GPU	CPU (Parallel)	CPU (Serial)
Read from disk	47.7	47.2	47.4
Preprocessing	2.9	2.9	2.9
MinHash	0.034	0.053	0.332
Join	145	2,988	27,964
Data transfer	0.53	0	0
Total	197	3,040	28,016

Table 3. Breakdown of the execution time in seconds when joining collections of the same size (Abstracts dataset, $|R| = |S| = 524,288$).

	GPU	CPU (Parallel)	CPU (Serial)
Read from disk	201.5	200.5	198.4
Preprocessing	9.3	9.4	9.1
MinHash	0.037	0.151	1.033
Join	145	1,403	11,955
Data transfer	0.09	0	0
Total	359	1,615	12,167

Table 4. Breakdown of the execution time in seconds when joining collections of the same size (Transactions dataset, $|R| = |S| = 524,288$).

	GPU	CPU (Parallel)	CPU (Serial)
Read from disk	379.8	378.4	376.2
Preprocessing	15.9	16.1	15.6
MinHash	0.040	0.250	1.728
Join	147	1,513	13,323
Data transfer	0.21	0	0
Total	549	1,914	13,723

5.4   Accuracy Evaluation

Since our scheme uses the MinHash technique to estimate the similarity between sets, it is also important to evaluate how accurate the results obtained from it are. We evaluated the accuracy of the proposal in terms of precision and recall. Precision relates to the fraction of really similar pairs among all the pairs retrieved by the algorithm, and recall refers to the fraction of really similar pairs that were correctly retrieved.

Table 5. Impact of varying number of bins on precision, recall and execution time (GPU implementation, Abstracts dataset, $|R| = |S| = 65,536$).

Number of bins	Precision	Recall	Execution time (s)
1	0.0000	0.9999	25.3
2	0.0275	0.9999	25.4
4	0.9733	0.9999	25.6
8	0.9994	0.9999	25.7
16	0.9998	1.0000	26.1
32	1.0000	1.0000	27.4
64	1.0000	1.0000	29.6
128	1.0000	1.0000	34.4
256	1.0000	1.0000	45.8
384	1.0000	1.0000	77.6
512	1.0000	1.0000	133.6
640	1.0000	1.0000	161.5

Table 5 presents the measurements of experiments in which we varied the number of bins composing the signatures of the sets, showing the impact of the number of bins on the number of similar pairs found, as well as on the performance.

Using a small number of bins (e.g., 1 or 2) results in dissimilar documents having similar signatures, thus making the algorithm retrieve a large number of pairs. Although most of the retrieved pairs are false positives (hence the low precision values), the majority of the really similar pairs is also retrieved, which is shown by the high values of recall. As the number of bins increases, the number of pairs retrieved nears the number of really similar pairs, thus increasing precision values.

On the other hand, increasing the number of bins also incurs a longer execution time. Therefore, it is important to achieve a balance between accuracy and execution time. For the used datasets, using 32 bins offered a good trade-off, yielding the lowest execution time without false positive or false negative results.

5.5   Other Experiments

We also conducted experiments varying other parameters of the implementation or characteristics of the data sets. For instance, Fig. 14 shows that, in the GPU implementation, varying the number of threads per block has little impact on the performance.

Figure 15 reveals that all three implementations are not significantly affected by varying the similarity threshold. In other words, although the number of similar pairs found changes, the GPU implementation is consistently faster than the other two.

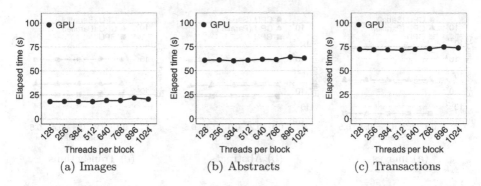

Fig. 14. Execution time varying the number of threads per GPU block ($|R| = |S| = 131,072$).

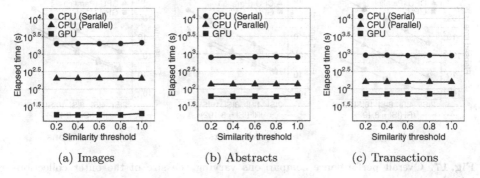

Fig. 15. Execution time varying the similarity threshold ($|R| = |S| = 131,072$).

Table 6. Precision and recall varying similarity threshold (GPU implementation, Abstracts dataset, $|R| = |S| = 8,192$).

Similarity threshold	Precision	Recall
0.2	0.08	0.89
0.4	0.99	0.99
0.6	1.00	1.00
0.8	1.00	1.00
1.0	1.00	1.00

Table 6 shows the impact of the similarity threshold on precision and recall levels. When the threshold is low, many pairs with a low degree of similarity are also part of the result (false positives). This situation is illustrated by the low precision value when the similarity threshold is 0.2.

Additionally, we constructed different collections of sets by varying the number of matching sets between them, i.e., the join selectivity. Figure 16 indicates that varying the selectivity does not impact the join performance.

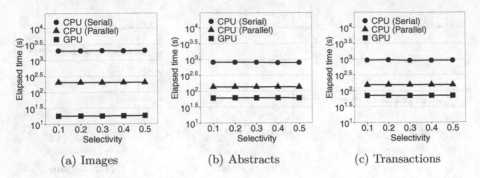

Fig. 16. Execution time varying the join minimum selectivity ($|R| = |S| = 131,072$).

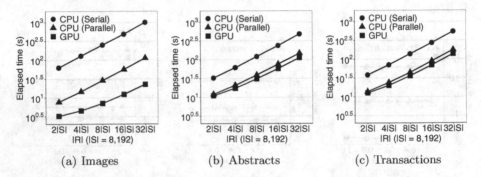

Fig. 17. Overall performance comparisons varying the size of the outer collection joined.

Finally, we also investigated the performance when joining collections of different sizes. Figure 17 shows the overall results when we vary the size of the collection processed the outer loop. Similar results were found when varying the size of the collection processed by the inner loop.

6   Related Work

This section presents works related to our proposal, which can be mainly divided in three categories: works that exploit GPUs for faster processing, works introducing novel similarity join algorithms, and works that, like ours, combine the previous two categories.

6.1   General-Purpose Processing on Graphics Processing Units

The use of GPUs for general processing is present in a number of areas nowadays (e.g., physics, chemistry and biology) [19]. In Computer Science, it has been used in network optimization [29], data mining [30], etc.

In a work that exploited GPU to perform relational joins, He et al. [7,8] presented implementations built on top of the parallel primitives mentioned previously. They evaluated the performance of four types of join (nested-loop join, indexed nested-loop join, sort-merge join and hash join) and obtained speedups of up to 7 times in comparison with the CPU counterpart. He et al. also considered relevant optimization aspects, like coalesced memory access to improve spatial locality, resulting in reduction of memory stalls and faster execution.

6.2 Similarity Joins

A survey done by Jiang et al. [11] made comparisons between a number of string similarity join approaches. The majority of these works focus on the elimination of unnecessary work and adopt a filter-verification approach [3,5,21,23,24,31–35], which initially prunes dissimilar pairs and leaves only candidate pairs that are later verified whether they are really similar. The evaluated algorithms were divided into categories, depending on the similarity metric they use. In the particular case of Jaccard similarity, AdaptJoin [23] and PPJoin+ [24] gave the best results. The survey included differences concerning the performance of algorithms based on the size of the dataset and on the length of the joined strings. Jiang et al. [11] also pointed out the necessity for disk-based algorithms to deal with really large datasets that do not fit in memory.

The adaptations of these serial algorithms for parallel environment can be seen as good opportunities for future work. Further investigation is necessary to determine if they are suitable for parallel processing, especially using GPUs, which require fewer memory transfers operations to be effective.

Other works focused on taking advantage of parallel processing to produce more scalable similarity join algorithms. Among these, Vernica et al. [20], Metwally et al. [16] and Deng et al. [4] used MapReduce to distribute the processing among nodes in CPU clusters.

6.3 GPU Accelerated Similarity Join

Although the similarity join is a thoroughly discussed topic, works utilizing GPUs for the processing speedup are not numerous. Lieberman et al. [15] mapped the similarity join operation to a sort-and-search problem and used well-known algorithms and primitives for GPUs to perform these tasks. After applying the bitonic sort algorithm to create a set of space-filling curves from one of the relations, they processed each record of the relation set in parallel, executing searches in the space-filling curves. The similarity between the records was calculated using the Minkowski metric.

Böhm et al. [1] proposed two GPU-accelerated nested-loop join (NLJ) algorithms to perform the similarity join operation, and used Euclidean distance to calculate the similarity in both cases. The best of the two methods was the index-supported similarity join, which has a preprocessing phase to create an index structure based on directories. The authors alleged that the GPU version of the indexed-supported similarity join achieved an improvement of 4.6 times when compared to its serial CPU version.

The main characteristic that discerns our work from the other similarity join schemes for GPUs is the effective use of MinHash to overcome challenges inherent to the use of GPUs for general-purpose computation, as emphasized in Sect. 2.2. Furthermore, to the best of our knowledge, our solution is the first one to couple Jaccard similarity and GPUs to tackle the similarity join problem.

A performance comparison with other works [1, 13, 15] was not possible since the source codes of previous solutions were not available.

7   Conclusions

We have proposed a GPU-accelerated similarity join scheme that uses MinHash in its similarity calculation step and achieved a speedup of more than two orders of magnitude when compared to the serial version of the algorithm. Moreover, the high levels of precision and recall obtained in the experimental evaluation confirmed the accuracy of our scheme.

The strongest point of GPUs is their superior throughput when compared to CPUs. However, they require special implementation techniques to minimize memory access and data transfer. For this purpose, using MinHash to estimate the similarity of sets is particularly beneficial, since it enables a parallelizable way to represent the sets in a compact manner, thus saving storage and reducing data transfer. Furthermore, our implementation explored the faster memories of GPUs (registers and shared memory) to diminish effects of memory stalls. We believe this solution can aid in the task of processing large datasets in a cost-effective way without ignoring the quality of the results.

Since the join is the most expensive part of the processing, future works will focus on the investigation and implementation of better join techniques on GPUs. For the algorithms developed in a next phase, the main requirements are parallelizable processing-intensive parts and infrequent memory transfers.

Acknowledgments. We thank the editors and the reviewers for their remarks and suggestions. This research was partly supported by the Grant-in-Aid for Scientific Research (B) (#26280037) from the Japan Society for the Promotion of Science.

References

1. Böhm, C., Noll, R., Plant, C., Zherdin, A.: Index-supported similarity join on graphics processors. BTW **144**, 57–66 (2009)
2. Broder, A.Z., Charikar, M., Frieze, A.M., Mitzenmacher, M.: Min-wise independent permutations. J. Comput. Syst. Sci. **60**(3), 630–659 (2000)
3. Chaudhuri, S., Ganti, V., Kaushik, R.: A primitive operator for similarity joins in data cleaning. In: Proceedings of ICDE, p. 5 (2006)
4. Deng, D., Li, G., Hao, S., Wang, J., Feng, J.: Massjoin: a MapReduce-based method for scalable string similarity joins. In: Proceedings of ICDE, pp. 340–351 (2014)

5. Gravano, L., Ipeirotis, P.G., Jagadish, H.V., Koudas, N., Muthukrishnan, S., Srivastava, D.: Approximate string joins in a database (almost) for free. In: Proceedings of VLDB, pp. 491–500 (2001)
6. Greathouse, J.L., Daga, M.: Efficient sparse matrix-vector multiplication on GPUs using the CSR storage format. In: Proceedings of SC, pp. 769–780 (2014)
7. He, B., Lu, M., Yang, K., Fang, R., Govindaraju, N.K., Luo, Q., Sander, P.V.: Relational query coprocessing on graphics processors. TODS **34**(4), 21:1–21:39 (2009)
8. He, B., Yang, K., Fang, R., Lu, M., Govindaraju, N., Luo, Q., Sander, P.: Relational joins on graphics processors. In: Proceedings of SIGMOD, pp. 511–524 (2008)
9. Hoberock, J., Bell, N.: Thrust: A Productivity-Oriented Library for CUDA. Morgan Kaufmann Publishers, San Francisco (2012)
10. Appleby, A.: MurmurHash3 (2016)
11. Jiang, Y., Li, G., Feng, J., Li, W.S.: String similarity joins: an experimental evaluation. PVLDB **7**(8), 625–636 (2014)
12. Li, P., Knig, A.C.: b-bit minwise hashing. CoRR abs/0910.3349 (2009)
13. Li, P., Shrivastava, A., König, A.C.: GPU-based minwise hashing. In: Proceedings of WWW, pp. 565–566 (2012)
14. Li, P., Owen, A.B., Zhang, C.H.: One permutation hashing for efficient search and learning. CoRR abs/1208.1259 (2012)
15. Lieberman, M.D., Sankaranarayanan, J., Samet, H.: A fast similarity join algorithm using graphics processing units. In: Proceedings of ICDE, pp. 1111–1120 (2008)
16. Metwally, A., Faloutsos, C.: V-Smart-Join: a scalable MapReduce framework for all-pair similarity joins of multisets and vectors. PVLDB **5**(8), 704–715 (2012)
17. NVIDIA Corporation: NVIDIA CUDA Compute Unified Device Architecture Programming Guide (2007)
18. OpenMP Architecture Review Board: OpenMP Application Program Interface Version 4.0 (2013)
19. Owens, J.D., Luebke, D., Govindaraju, N., Harris, M., Krger, J., Lefohn, A., Purcell, T.J.: A survey of general-purpose computation on graphics hardware. Comput. Graph. Forum **26**(1), 80–113 (2007)
20. Rares, V., Carey, M.J., Chen, L.: Efficient parallel set-similarity joins using MapReduce. In: Proceedings of SIGMOD, pp. 495–506 (2010)
21. Sarawagi, S., Kirpal, A.: Efficient set joins on similarity predicates. In: Proceedings of SIGMOD, pp. 743–754 (2004)
22. Sengupta, S., Harris, M., Zhang, Y., Owens, J.D.: Scan primitives for GPU computing. In: Proceedings of GH, pp. 97–106 (2007)
23. Wang, J., Li, G., Feng, J.: Can we beat the prefix filtering? An adaptive framework for similarity join and search. In: Proceedings of SIGMOD, pp. 85–96 (2012)
24. Xiao, C., Wang, W., Lin, X., Yu, J.X.: Efficient similarity joins for near duplicate detection. In: Proceedings of WWW, pp. 131–140 (2008)
25. Cruz, M.S.H., Kozawa, Y., Amagasa, T., Kitagawa, H.: GPU acceleration of set similarity joins. In: Chen, Q., Hameurlain, A., Toumani, F., Wagner, R., Decker, H. (eds.) DEXA 2015. LNCS, vol. 9261, pp. 384–398. Springer, Heidelberg (2015)
26. Harris, M.: Parallel prefix sum (Scan) with CUDA (2009)
27. Dotsenko, Y., Govindaraju, N.K., Sloan, P., Boyd, C., Manferdelli, J.: Fast scan algorithms on graphics processors. In: Proceedings of ICS, pp. 205–213 (2008)
28. Yan, S., Long, G., Zhang, Y.: StreamScan: fast scan algorithms for GPUs without global barrier synchronization. In: Proceedings of PPoPP, pp. 229–238 (2013)
29. Han, S., Jang, K., Park, K., Moon, S.: PacketShader: a GPU-accelerated software router. In: Proceedings of SIGCOMM, pp. 195–206 (2010)

30. Gainaru, A., Slusanschi, E., Trausan-Matu, S.: Mapping data mining algorithms on a GPU architecture: a study. In: Kryszkiewicz, M., Rybinski, H., Skowron, A., Raś, Z.W. (eds.) ISMIS 2011. LNCS, vol. 6804, pp. 102–112. Springer, Heidelberg (2011)
31. Li, G., Deng, D., Wang, J., Feng, J.: Pass-join: a partition-based method for similarity joins. PVLDB **5**, 253–264 (2011)
32. Xiao, C., Wang, W., Lin, X.: Ed-Join: an efficient algorithm for similarity joins with edit distance constraints. PVLDB **1**, 933–944 (2008)
33. Bayardo, R., Ma, Y., Srikant, R.: Scaling up all pairs similarity search. In: Proceedings of WWW, pp. 131–140 (2007)
34. Ribeiro, L., Härder, T.: Generalizing prefix filtering to improve set similarity joins. Inf. Syst. **36**, 62–78 (2011)
35. Wang, W., Qin, J., Chuan, X., Lin, X., Shen, H.: VChunkJoin: an efficient algorithm for edit similarity joins. TKDE **25**, 1916–1929 (2013)

Divide-and-Conquer Parallelism
for Learning Mixture Models

Takaya Kawakatsu[1]([✉]), Akira Kinoshita[1], Atsuhiro Takasu[2],
and Jun Adachi[2]

[1] The University of Tokyo, 2-1-2 Hitotsubashi, Chiyoda, Tokyo, Japan
{kat,kinoshita}@nii.ac.jp
[2] National Institute of Informatics, 2-1-2 Hitotsubashi, Chiyoda, Tokyo, Japan
{takasu,adachi}@nii.ac.jp

Abstract. From the viewpoint of load balancing among processors, the
acceleration of machine-learning algorithms by using parallel loops is not
realistic for some models involving hierarchical parameter estimation.
There are also other serious issues such as memory access speed and
race conditions. Some approaches to the race condition problem, such
as mutual exclusion and atomic operations, degrade the memory access
performance. Another issue is that the first-in-first-out (FIFO) scheduler
supported by frameworks such as Hadoop can waste considerable time
on queuing and this will also affect the learning speed. In this paper, we
propose a recursive divide-and-conquer-based parallelization method for
high-speed machine learning. Our approach exploits a tree structure for
recursive tasks, which enables effective load balancing. Race conditions
are also avoided, without slowing down the memory access, by separating
the variables for summation. We have applied our approach to tasks
that involve learning mixture models. Our experimental results show
scalability superior to FIFO scheduling with an atomic-based solution to
race conditions and robustness against load imbalance.

Keywords: Divide and conquer · Machine learning · Parallelization ·
NUMA

1 Introduction

There is growing interest in the mining of huge datasets against a backdrop
of inexpensive, high-performance parallel computation environments, such as
shared-memory machines and distributed-memory clusters. Fortunately, modern
computers can have large memories, with hundreds of gigabytes per CPU socket,
and the memory size limitation may not continue to be a severe problem in itself.
For this reason, state-of-the-art parallel computing frameworks like Spark [1,2],
Piccolo [3], and Spartan [4] can take an *in-memory* approach that stores data in
dynamic random access memory (DRAM) instead of on hard disks. Nonetheless,
there remain four critical issues to consider: *memory access speed*, *load imbalance*,
race conditions, and *scheduling overhead*.

© Springer-Verlag Berlin Heidelberg 2016
A. Hameurlain et al. (Eds.): TLDKS XXVIII, LNCS 9940, pp. 23–47, 2016.
DOI: 10.1007/978-3-662-53455-7_2

A processor accesses data in its memory via a bus and spends considerable time simply waiting for a response from the memory. In shared-memory systems, many processors can share the same bus. Therefore, the latency and throughput of the bus will have a great impact on calculation speed. For distributed-memory systems in particular, each computation node must exchange data for processing via message-passing frameworks such as MPI[1], with even poorer throughput and greater latency than bus-based systems. Therefore, we should carefully consider memory access speeds when considering the computation speed of a program. The essential requirement is to improve the reference locality of the program.

Load imbalance refers to the condition where one processor can be working hard while another processor is waiting idly, which can cause serious throughput degradation. In some data-mining models, the computation cost per observation data item is not uniform and load imbalance may occur. To avoid this, dynamic scheduling may be a solution.

Another characteristic issue in parallel computation is the possibility of race conditions. For shared-memory systems, if several processors attempt to access the same memory address at the same time, the integrity of the calculation can be compromised. Mutual exclusion using a semaphore [5] or mutex can avoid race conditions, but can involve substantial overheads. As an alternative, we can use atomic operations supported by the hardware. However, this may remain expensive because of latency in the cache-coherence protocol, as discussed later.

The fourth issue is scheduling overhead. The classic first-in-first-out (FIFO) scheduler supported by existing frameworks such as OpenMP[2] and Hadoop[3] is implemented under a *flat partitioning* strategy, which divides and allocates tasks to each processor without detailed consideration of their interrelationships. A flat scheduler cannot adjust the granularity of the subtasks and it tends to allocate tasks with extremely small granularity. Because a FIFO scheduler has only one task queue and all processors access the queue frequently, the queuing time may become a serious bottleneck, particularly with fine-grained parallelization.

In this paper, we propose a solution for these four issues by bringing together two relevant concepts: *work-stealing* [6,7] and the *buffering solution* under a recursive divide-and-conquer-based parallelization approach called *ADCA*. The combination of a work-stealing scheduler with our ADCA will reduce scheduling overheads because of the absence of bottlenecks, while ADCA also achieves efficient load balancing with optimum granularity. Buffering is a method whereby each processor does local calculations wherever possible, with a master processor integrating the local results later. This helps to avoid both race conditions and latency caused by the cache-coherence protocol. ADCA and the buffering solution are our main contributions.

As target applications for ADCA, we focus on machine-learning algorithms that repeat a learning step many times, with each step handling the observation data in parallel. Expectation-maximization (EM) algorithms [8,9] on

---

[1] http://www.mpi-forum.org.

[2] http://www.openmp.org.

[3] http://hadoop.apache.org.

a Gaussian mixture model (GMM) or a hierarchical Poisson mixture model (HPMM) [10,11] are well-known examples of such applications. Mixture models are popular and versatile; their applications include wireless sensor networks [12,13], speech recognition [14,15], and moving object detection [16–19]. Another principal application, back-propagation-based learning [20] of neural networks, can also be parallelized using the same approach [21,22].

In Sect. 2, we formulate parallel computing in general terms, introducing our main concept, *three-step parallel computing*, and then introduce work-stealing and the buffering solution. In Sect. 3, we summarize related work on parallel EM algorithms and then explain our EM algorithm based on ADCA. In Sect. 4, we demonstrate our method's superior scalability to FIFO scheduling and to the atomic solution by experiments with GMMs. We also demonstrate our method's robustness against load imbalance by experiments with HPMMs. Finally, we conclude this paper in Sect. 5.

2   Parallel Computation Models

There are a vast number of approaches to parallel computing; it is not easy for users to select an approach that meets their requirements. Even though parallel technologies may not seem to cooperate with each other, we can integrate them according to the *three-step parallel computing* principle, which contains three phases: *parallelization*, *execution*, and *communication*. In the parallelization phase, the programmer writes the source code specifying those parts where parallel processing is possible. In the execution phase, a computer executes the program serially, assigning tasks to its computation units as required. Finally, in the communication phase, the units synchronize and exchange values.

2.1   Parallelization of Algorithms

In the parallelization phase, the programmer effectively informs the computer which statements can be executed in parallel. This can be separated into two subphases: the *algorithm phase* and the *directive phase*. In the algorithm phase, the programmer chooses the form of parallelism: *data parallelism* [23] or *task parallelism*. The programmer then specifies the parallelizable statements in the directive phase. For data parallelism, the program is described as a loop, as illustrated in Fig. 1a. That is also called *loop parallelism*. OpenMP supports loop parallelism by the directive `parallel for`. Furthermore, single-instruction multiple-data (SIMD) [24] instructions, such as Intel streaming SIMD extensions, can be categorized as data parallelism. When exploiting data parallelism, we must assume that the program describes an operator that takes an array element as an argument.

Next, task parallelism can be described by using `fork` and `join` functions in a recursive manner, as illustrated in Fig. 1b. After the `fork` function is called, a new thread is created and executed. Each thread processes a user-defined task, and the calculation result is returned to the invoker thread by calling the `join`

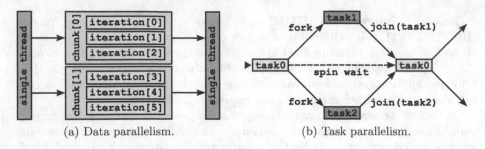

(a) Data parallelism.   (b) Task parallelism.

Fig. 1. Data parallelism and task parallelism. A parallel program can be described in a loop manner or a fork–join manner.

function. Actually, the `fork` and `join` functions are provided by the pthreads, `pthread_create` and `pthread_join`, respectively.

In many cases, the critical statement that has the most significant impact on the execution time is a `for` loop with many iterations. A data-parallel program can be much simpler than a task-parallel program. For this reason, parallel loops are frequently exploited in computationally heavy programs. The EM algorithm on a GMM can be parallelized in the loop manner [25–30]. However, parallel loops are not applicable when the data have mostly nonarray structures like graphs or trees. The HPMM is a simple example of such a case. Therefore, parallelizable machine learning for graphical models must be described in a *fork–join* manner.

In practice, data and task parallelism can work together in a single program, such as forking tasks in a parallel loop or exploiting a parallel loop in a recursive task, because parallel loops can be treated as the syntactical sugar of the `fork` and `join` functions. Of course, there are devices that hardly support task parallelism, such as graphical processing units (GPUs). Task parallelism on a GPU remains a challenging problem [31,32].

Finally, the directive phase can be categorized as involving *explicit directives* or *implicit directives*. The `fork` and `join` functions are examples of explicit directives that permit programmers to describe precisely the relationships between forked tasks. For the implicit case, a scheduler determines automatically whether statements are to be executed in parallel or serially. That decision is realized on the assumption that each task has referential transparency. That is, there are no side effects such as destructive assignment to a global variable.

2.2   Parallel Execution Mechanism

A program that manages tasks is called a *scheduler*, dealing with three subphases: *traversal*, *delivery*, and *balancing*.

In the traversal phase, the scheduler scans the remaining tasks to determine the order of task execution. In task parallelism, tasks have a recursive tree-based structure, and in general, there are two primary options, depth-first traversal, or breadth-first traversal.

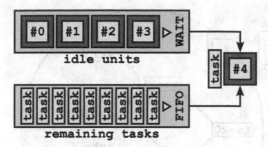

Fig. 2. General FIFO-based solution to counter load imbalance. The program is partitioned into tasks that are allocated one by one to the computation units.

Then, in the delivery phase, the scheduler determines the computation unit that executes each task. This phase plays an important role in controlling reference locality, with the scheduler aiming to reduce load-and-store latency by allocating each task to a computation unit located near the data associated with that task. This is particularly important for machine-learning algorithms, where the computer must repeat a learning step many times until the model converges. Ideally, the scheduler should assign tasks to the computation units so that each unit handles the same data chunk in every learning step, reducing the necessity for data exchanges between units. However, such an optimization does not make sense if the program then has serious load imbalances. In some machine-learning algorithms, the computation cost per task may not be uniform.

In the balancing phase, the scheduler relieves a load imbalance when it detects an idling computation unit. This is an ex post effort, whereas the delivery phase is an ex ante effort. There are two options for this phase: *pushing* [33–35] and *pulling* [6,7,36–40]. Pushing is when a busy unit takes the initiative as a producer and sends its remaining tasks to idling units by passing messages whenever requested by the idling units. In contrast, pulling is when an idle unit takes the initiative as a consumer and snatches its next task from another unit.

The FIFO scheduling illustrated in Fig. 2 is a typical example of a pulling scheduler. An idling unit tries to snatch its next task for execution from a shared task queue called the *runqueue*. The program is partitioned into many subtasks that are appended to the runqueue by calling the `fork` function. Because of its simplicity, FIFO scheduling is widely used in Hadoop and UNIX. While this may appear to be a good solution, it can cause excessively fine-grained task snatching and the resulting overhead will reduce the benefits of the parallel computation. A shared queue is accessed frequently by all computation units and therefore behaves as a single point of failure. Hence, the queuing time may become significant, even though the queue implementation utilizes a lock-free-based protection technique instead of mutual exclusion such as a mutex. To avoid this, the task partitioning should be as coarse-grained as possible; however, load balancing will then be less effective. As another issue, we suspect that the ability

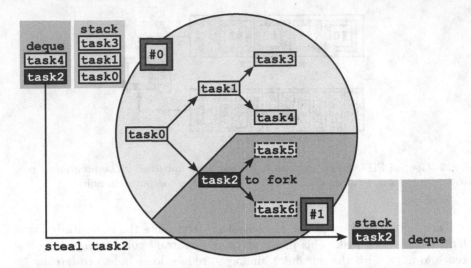

Fig. 3. Breadth-first task distribution with a work-stealing scheduler. Idle unit #1 steals a task in a FIFO fashion to minimize the stealing frequency.

to tune referential locality can be poor and this will become a serious problem, particularly in the context of distributed processing.

Mohr et al. introduced a novel balancing technique called *work-stealing* for their LISP system [6, 41]. They focused on the property of a recursive program that the size of each task can be halved by expanding the tree-structured tasks. As illustrated in Fig. 3, the work-stealing scheduler first expands the root task into a minimum number of subtasks, and distributes them to computation units either by pushing or pulling. When a computation unit becomes idle, the scheduler divides another unit's task in half and reassigns one half to the idle unit. This behavior is called work-stealing. In this way, the program is always divided into the smallest number of tasks required, thereby achieving a minimum number of stealing events.

A typical work-stealing scheduler [35, 40, 42] is constructed by exploiting a thread library such as pthreads. Each computation unit is expressed as a *worker thread* fixed to the unit and has its own local *deque* or double-ended queue to hold tasks remaining to be executed. Tasks are popped by the owner unit and executed one by one in a last-in first-out (LIFO) fashion. When there are no idling units, each worker behaves independently of the others. If a unit becomes idle, with an empty deque, the unit scouts around other units' deques until it finds a task and steals it in a FIFO fashion, as described in Algorithm 1.

Of course, a remaining task may create new tasks by calling a `fork` function, and such subtasks are appended into the local deque in a LIFO fashion, as shown in Algorithm 1. Hence, the tasks in each deque are stored in order of descending age. That is, an idling unit steals the oldest remaining task, and that will be the one nearest to the root of the task tree. This is the reason for the work-stealing

---

Algorithm 1. A worker thread's behavior in a work-stealing scheduler.

Require: *myself*: the worker thread, *victim*: another worker thread
  procedure FORK(*function, arguments*)
    task = new task(*function, arguments*)
    myself.deque.append(*task*)
    return *task*
  end procedure
  procedure JOIN(*task*)
    repeat
      if *myself*.deque.is_empty **then**
        next = *victim*.deque.pop_FIFO()
        next.execute()
      else
        next = *myself*.deque.pop_LIFO()
        next.execute()
      end if
    until *task*.is_finished
  end procedure

---

scheduler being able to achieve the minimum number of stealing events necessary. In addition, there is no single point of failure, and the overhead will be smaller than that for the FIFO scheduler.

2.3 Communication Mechanism

In the communication mechanism, each computation unit exchanges values via a bus or network. For example, when calculating the mean value of a series, each unit will calculate the mean of a chunk, with a master unit then unifying the means into a single value. There are two options for the communication mechanism: *distributed memory* and *shared memory*.

In a distributed-memory system, each computation unit has its own memory and its address space is not shared with other units. Communication among units is realized by explicit message passing, with greater latency than local memory access.

In a shared-memory system, several computation units have access to a large memory with an address space shared among all computation units. There is no need for explicit message passing, with communication achieved by reading from or writing to shared variables.

A great problem for shared-memory systems is the possibility of *race conditions*, as shown in Fig. 4a. Suppose that two computation units, #0 and #1, are adding some numbers into a shared variable `total` concurrently. Such an operation is called a *load-modify-store* operation, but the result can be incorrect, because of conflicts between loading and storing. Using atomic operations can be a solution. An atomic operation is guaranteed to exclude any load or store operations by other computation units until the operation finishes.

Note that, because memory access latency suspends an operation for a time, modern processors support the *out-of-order* execution paradigm, going on to

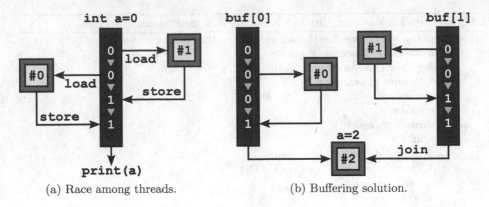

(a) Race among threads.   (b) Buffering solution.

Fig. 4. A race condition among computation units and our buffering solution. The sum may be incorrect if several units access the same variable at the same time.

execute other instructions until the processor obtains all required data from the memory. This accelerates serial program execution, but can compromise the integrity of a parallel program that includes some critical instructions that must be strictly executed in a particular order. Figure 5 is an example of such a program, where `thread2` waits until `thread1` updates `value`. The loop is called *spin waiting* or *busy waiting*, and is frequently used for thread synchronization. If the statements are executed out of order, `thread2` may load an old value before the update by `thread1`. As a solution, an atomic operation often involves a *memory fence*, instructions that inhibit out-of-order execution.

Considering the memory access patterns of machine-learning algorithms, the use of atomic operations may not be an adequate solution. Main memories based on DRAM operate more slowly than processors, so that modern computers insert caches based on static random-access memory (SRAM) between the processors and main memories. Main memory access is blocked if the cache memory has a valid replica of the addressed data. The problem is that several computation units may have replicas of the same data item in their own cache memories and the stored replica values may become outdated. Machine-learning algorithms access all the observation items simultaneously to calculate summations in each learning step, and cache conflicts may occur frequently. A *cache-coherence* [43, 44] protocol invalidates the old replica to relieve conflicts, as illustrated in Fig. 6. However, this may become a serious bottleneck. For this reason, we recommend a *buffering solution*, as shown in Fig. 4b. The solution separates the memory addresses physically to avoid cache conflicts. Each computation unit calculates a summation into its local buffer and the master unit retrieves these to calculate the total summation after calling a `join` function. This method can be combined easily with our divide-and-conquer-based parallelization approach.

2.4   Parallel Computing Frameworks

Many frameworks assist parallelization, parallel execution, and communication.

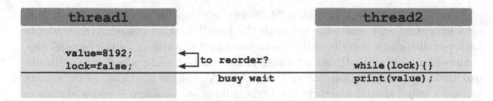

thread1		thread2
value=8192; lock=false;	to reorder? busy wait	while(lock){} print(value);

Fig. 5. Data exchange between threads by spin waiting. The synchronization will fail if the statements are executed out of order.

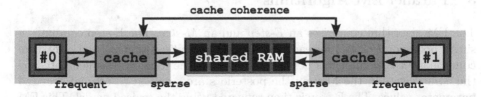

Fig. 6. Cache coherence among computation units. The protocol will cause a bottleneck whenever units are accessing the same address.

As a parallelization framework, we could use a high-performance-computing (HPC) language such as Cilk [7,36], X10 [45], or Chapel [46]. They support task parallelism, and Chapel also supports data parallelism with sophisticated syntax. OpenMP is another well-known framework for task and data parallelism for shared-memory environments. These languages and libraries are examples of the explicit-directive approach, while MapReduce [47] is a library-level variant of the implicit-directive approach.

As a parallel execution framework, we could use a work-stealing scheduler such as Intel's TBB[4], qthreads [38,39], or MassiveThreads [40]. Pthreads[5] does not support work-stealing by itself, but is an essential component for implementing such schedulers.

In the context of communication mechanisms, some general-purpose languages provide helpful programming models for parallel computation, even though they were not designed primarily as HPC languages. For example, Go[6] supports sophisticated syntax for message passing. In low-level programming, MPI is a standardized message-passing framework that supports MPI_Send and MPI_Recv. In the context of enterprise applications, Hadoop is a well-maintained platform for distributed data processing. In the context of shared-memory communication, the simplest example is multithread programming using a thread library. In C++11, all global variables are shared among threads by default, unless using a thread_local specifier, and by using std::atomic templates[7], a programmer can write thread-safe access to shared data.

---

[4] http://www.threadingbuildingblocks.org.

[5] http://computing.llnl.gov/tutorials/pthreads/.

[6] http://golang.org.

[7] http://www.cplusplus.com/reference/atomic/atomic.

The ADCA proposal in Sect. 3.2 follows the principles of parallelization, parallel execution and communication. In the parallelization phase, ADCA adopts task parallelism to describe a divide-and-conquer algorithm. In the parallel execution phase, ADCA utilizes a pulling-based work-stealing scheduler to distribute tasks to computation units. Then, ADCA realizes communication among units by using shared-memory; the cache coherence problem is reduced thanks to the buffering solution.

3   Parallel EM Algorithms

The EM algorithm comprises an *E-step* and an *M-step*. The E-step computes a single posterior $P(k|x_n)$ for each pair of an observation item x_n and a mixture component k that indicates how likely it is that the item x_n was generated by the component. In the M-step, the posteriors are summed to estimate revised parameter values. The E-step is then repeated using the revised model. This EM iteration continues until the likelihood function $\mathcal{L}$, which indicates how well the model regenerates the dataset, converges to a maximum.

To parallelize the EM algorithm, we should divide each E-step and M-step into a number of tasks and allocate them to computation units. For GMMs, the axis of the observation item x_n and the axis of mixtures k are available for this division. As illustrated in Fig. 7, reference locality is maximized whenever we divide the posterior table into squares because the total number of observation items and model parameters to be loaded to calculate the posterior subtable is minimized. The number of observation items is usually much greater than the number of mixture components. Consequently, we must divide the n axis more than the k axis.

The EM algorithm includes summation processes in the M-step, and we must take measures against race conditions among computation units. In general, race conditions are resolved by using mutual exclusion techniques such as semaphores, or by using atomic instructions. However, we do not recommend these approaches because mutual exclusion involves a blocking time, and an atomic operation on a shared variable involves the cache-coherence problem. The best solution is to let each computation unit use its own local memory to hold intermediate results, thereby computing a partial sum independently, before a single unit retrieves the intermediate results to compute the total.

3.1   Related Work

Nonuniform memory access (NUMA) is a shared-memory architecture for which a computation unit and local memory form a pair called a *NUMA node*. Modern processors support NUMA at the chip level and a NUMA node is generally equivalent to a processor socket. The memory address space is continuous, enabling each NUMA node to access another node's local memory in the same way as for its own local memory. In a shared-memory system, all computation units share the threads of a process and any unit can execute a thread. Because the threads

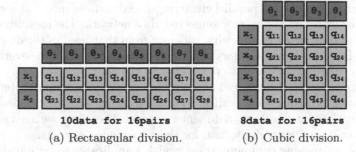

10data for 16pairs   **8data for 16pairs**

(a) Rectangular division.   (b) Cubic division.

Fig. 7. Cubic division of the posterior table. Space requirements are minimized when the posterior table is divided into cubes. The symbols such as θ and q_{nk} are described in detail in Appendix A.

share a common address space assigned to the process, we can implement programs without explicit message passing among threads. However, noting that all NUMA nodes will be interconnected via a bus, nonlocal memory accesses will have higher latency than local memory accesses for a computation unit. Consequently, programmers should aim to maximize reference locality to increase the proportion of local memory accesses. Kwedlo [25] proposed a parallel version of the EM algorithm for a NUMA computer. He used the parallel loop of OpenMP and introduced two techniques for improving the reference locality, the buffering solution and *first touching*.

When we parallelize an EM algorithm, we normally partition the observation items into a number of data chunks and assign them to threads. On the one hand, a race condition will never arise in the E-step because there is no summation over the observation items. On the other hand, parameter recalculation in the M-step requires summation over the posteriors. Whenever several computation units read from and write to the same address *simultaneously*, a race condition is possible, with the revised parameters differing from the correct values. To avoid this, Kwedlo arranged an independent array for each unit, with each unit calculating a partial sum into its own array. The partial sums are then integrated by a single thread at the end of the M-step. He introduced a buffering solution, although OpenMP supports safe summation via the `reduction` clause, because that clause cannot handle array types [25]. Accordingly, Kwedlo proposed his own buffering solution, coincidentally similar to Fig. 4b.

First-touching [48,49] is a well-known optimization method in the context of combining Linux[8] and NUMA. In Linux, a logical memory address is not bound to a physical memory address initially. When a thread accesses the logical address for the first time, a small physical memory space called a *page*, is selected from the closest NUMA node and allocated to the logical address. Therefore, Kwedlo made all threads access their own chunk before running the EM algorithm [25].

---

[8] https://www.kernel.org.

Distributed memory is a parallel clustering-based architecture that comprises many computers called *nodes* interconnected via a network. The meaning of *node* in distributed computing is somewhat different from that used in the description of NUMA. For distributed memory, each node is a processor with several cores that use a shared-memory architecture. That is, the distributed-memory system has at least two levels of memory hierarchy: *internode* and *intranode*. Internode communication is realized by explicit message passing, with internode latency being greater than intranode shared-memory access. Therefore, we must consider reference locality more carefully than for shared-memory systems.

The message-passing communication model is applicable to both distributed and shared-memory systems. It rarely depends on the detailed architecture, with its application being wider than that of the shared-memory programming model. For this reason, the message-passing model is more popular with programmers using high-level languages such as Java and Scala. A programmer can use highly abstract concurrent-execution models such as MapReduce, with the background scheduler then assigning tasks to the computation units. MapReduce, supported by Hadoop, offers a simple but powerful abstraction. However, that is too abstract to enable control of the reference locality, unlike NUMA programming exploiting the first-touch policy. Therefore, MapReduce might be convenient but can result in poor throughput. Its handling of hard-disk I/O overhead is another principal reason for its poor performance [1].

Currently, there are several parallel-computing frameworks based on message passing, such as Spark [1], Piccolo [3], and GraphLab [50,51]. Spark is a framework that aligns data in memory to reduce hard-disk I/O overheads, and provides its own distributed, immutable collection framework called the resilient distributed dataset (RDD) [2]. Because RDD elements are in memory, Spark runs faster than Hadoop MapReduce, which must read observation dataset from hard disks each time they are required. Piccolo is a distributed in-memory hash-table framework that runs parallel applications with high efficiency, similarly to RDD. GraphLab is a distributed machine-learning framework in which the programmer describes calculations and data flows by using directed graphs. Each node behaves as if it was a local Map or Reduce facility, with the many Map and Reduce operations all running in parallel.

In the world of low-level programming, *hybrid parallel computing* [52–54] is a popular approach. It uses a thread implementation such as pthreads inside the nodes and MPI among the nodes. Yang et al. [26] proposed an EM algorithm using a hybrid parallelization approach. It divides and conquers the observation items and integrates partial sums at the end of the M-step, as shown in Fig. 8. The implementation has a hierarchical structure. First, the master node assigns an observation data subset to each distributed node by utilizing MPI. Next, each distributed node partitions its subset into smaller subsets and allocates them to threads running on each node. At the end of every M-step, each distributed node calculates its internal summation and the master node collects them to calculate the total sum. This approach achieves good reference locality because nodes do not exchange any values until the intranode calculations are completely finished.

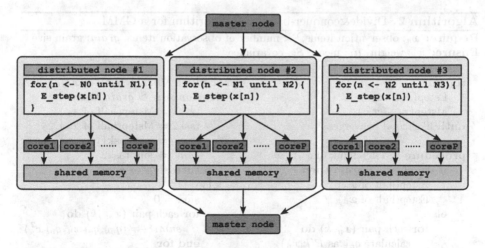

Fig. 8. Hybrid message-passing and shared-memory parallelization. This approach uses a thread inside the node and message passing among the nodes.

The authors attribute the reduction in scheduling overheads to *static scheduling*. That is, their approach divides the data into equal-sized subsets before running the EM algorithm and never deals with load balancing. Of course, the processing throughput might suffer if load imbalances occur.

3.2 Our ADCA Proposal

To utilize a work-stealing scheduler, we must transform the EM algorithm to a divide-and-conquer form, which is not difficult because we can divide the observation dataset recursively and calculate the posteriors in parallel in the E-step. In the M-step, we repartition the dataset to recalculate the parameters recursively as shown in Algorithm 2. This version of the EM algorithm performs effective dynamic load balancing.

ADCA also achieves good reference locality. The divided observation subset and its counterpart will be aligned closely in the memory address space. Machine-learning algorithms repeat learning steps until the model regenerates the training data. Therefore, the programmer can optimize ADCA so that each computation unit retrieves a chunk of the dataset from storage before processing, and handles only its own local chunk in every subsequent step. This optimization may reduce internode I/O transactions dramatically in distributed-memory systems, which is a direction for our future work.

Note that the observation items should not be divided into single data items because the calculation cost per observation item would then be very small, and too-frequent task-switching operations might degrade the processing throughput. Therefore, we introduce a *grain size* parameter as the minimum size for a subset of the observation data. When recursive division of the subsets reaches the grain size, no further division occurs. Although we do not examine selection methods

Algorithm 2. Divide&conquer-based EM algorithm for a GMM.

Require: x_n: observation items, N: number of observation items, $grain$: grain size
Ensure: w_k: weight, μ_k: mean, S_k: covariance

```
repeat                                procedure MSTEP(chunk of x_snk)
    Estep(x_1, .., x_N)                  if chunksize > grain then
    Mstep(x_1, .., x_N)                      task1 = Mstep(half of x_n)
until likelihood converges                   task2 = Mstep(half of x_n)
                                             sum^1 = join task1
procedure ESTEP(chunk of x_n)                sum^2 = join task2
    if chunksize > grain then                return sum^1 + sum^2
        Estep(half of x_n)               else
        Estep(half of x_n)                   sum = 0
    else                                     for each pair (x_sn, k) do
        for each pair (x_n, k) do                sum_k += (q_nk, q_nk x_n, q_nk x_n^2)
            calculate q_nk as P(k|x_n)       end for
        end for                              return sum
    end if                               end if
end procedure                         end procedure
```

for the grain size here, it can be smaller than for the FIFO scheduler because of the small overhead of the work-stealing-based scheduler, as described in Sect. 2.2.

ADCA has another advantage, the avoidance of race conditions. As shown in Algorithm 2, the parameter recalculation in the M-step is implemented using buffering, and there are no critical sections. Accordingly, there is no need for mutual exclusion or atomic operations on shared variables, which enables much faster computation. Of course, our approach may have the disadvantage of requiring more memory than other approaches do.

4   Experiment and Results

Table 1 describes the experimental environments. For most experiments, we used the hu080, but the strong and weak scaling of ADCA were also measured using the hp160. Both are shared-memory computers with many NUMA nodes, using CentOS[9] and the GNU compiler collections (gcc[10]) 5.2 and 4.8.

To demonstrate the scalability and robustness against load imbalance of our approach, we transformed the EM algorithms into the divide-and-conquer form shown in Algorithm 2. In a previous paper [55], we described the implementation of ADCA in the programming language Chapel, employing MassiveThreads [40] as the work-stealing scheduler, and compared it with a FIFO approach that used OpenMP. For the purposes of this paper, we have implemented a work-stealing scheduler and a FIFO scheduler in C++11 to unify the experimental conditions,

---

[9] http://www.centos.org.
[10] http://gcc.gnu.org.

Table 1. Experimental machine environments.

Name	hu080		hp160	
CPU	Xeon	E7 4870	Xeon	E7 8891 v2
	Clock	2.4 GHz	Clock	3.2 GHz
	Cores	10	Cores	10
Cache	L1d	32 kB/core	L1d	32 kB/core
	L2	256 kB/core	L2	256 kB/core
	L3	30.0 MB	L3	37.5 MB
NUMA	Nodes	8	Nodes	16
	RAM	64 GB/node	RAM	0.75 TB/node

excluding the schedulers and parallelization approaches. Both schedulers employ the lock-free deque implementation proposed by Arora et al. [56], instead of using a mutex-based deque such as MassiveThreads. We examined three aspects of the system: the effect of the buffering solution, robustness against fine-grained parallelism, and robustness against load imbalance. The effect of the buffering solution was ascertained by comparison with the atomic solution using a GMM. Robustness against fine-grained parallelism was tested by comparison with the FIFO approach, again using the EM algorithm on a GMM. Robustness against load imbalance was examined by learning both load-imbalanced and load-balanced HPMM datasets.

For the experiments, we prepared randomly generated training data. For the GMMs, each observation item was generated by eight mixture components, and was expressed as an eight-dimensional vector, with each value expressed in 64-bit floating-point form. For the HPMMs, each item was generated by eight mixture components, and was expressed as a two-dimensional 64-bit integer vector.

The graphs below show results for both *strong* and *weak scaling*. They indicate how processing speed varies with the number of computation units, but the size of the dataset is fixed in strong scaling, whereas the size varies in proportion to the number of units in weak scaling. For strong scaling, the vertical axis indicates the throughput in megarecords per second (MRPS), whereas it indicates the processing time per EM iteration for weak scaling.

4.1 Effect of Buffering Solution

To demonstrate the effect of the buffering solution, we compared the difference in scalability between the buffering solution and the atomic solution for the EM algorithm on a GMM. We adopted the FIFO approach in both cases and set the grain size to 256. That is, each task handles 256 observation items. The datasets had 268,435,456 items in total in strong scaling and 2,097,152 items per core in weak scaling. Figure 9 shows the comparison results. The atomic solution did not so much speed up as slow down in weak scaling, whereas the buffering solution achieved an almost linear speedup. The atomic solution decelerated at a rate

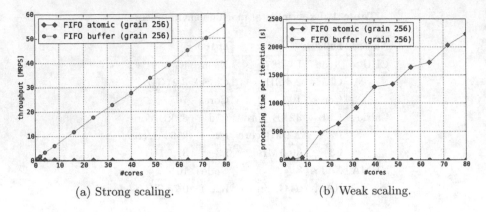

(a) Strong scaling.   (b) Weak scaling.

Fig. 9. Atomic solution vs buffering solution (hu080).

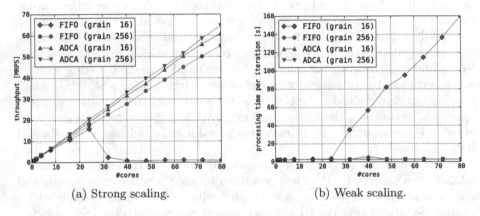

(a) Strong scaling.   (b) Weak scaling.

Fig. 10. Scalability of the EM algorithm on a GMM (hu080).

of 28 seconds per core. Note that the total throughput was invariant regardless of the number of cores. This suggests that there exists a bottleneck setting the upper limit of the throughput. We suspect that the cache-coherence protocol was the main factor.

4.2 Robustness Against Fine-Grained Parallelism

To demonstrate the robustness against fine-grained parallelization, we compared the difference in scalability between the FIFO-based approach and ADCA, while varying the grain size from 256 items to 16 items. The sizes of the datasets were the same as those for Fig. 9. Figure 10 shows the evaluation result. When the grain size was set to 256, the FIFO approach accelerated between 1 and 80 cores at a rate that was 15.7 % less than that for ADCA. When the grain size was set to 16, the FIFO approach decelerated beyond 24 cores, and it did not speed up beyond a factor of 17.5. In contrast, our approach achieved a near-linear speedup in both cases. This could be explained by the overhead of the shared runqueue.

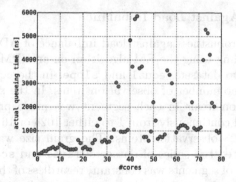

Fig. 11. Maximum queuing time with FIFO scheduling (hu080).

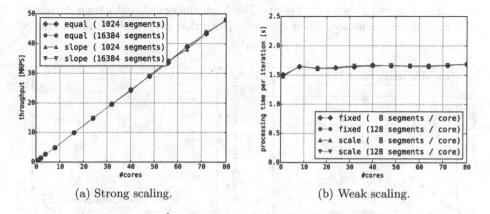

(a) Strong scaling.

(b) Weak scaling.

Fig. 12. Scalability of the EM algorithm on an HPMM (hu080).

To test that hypothesis, we also measured the queuing time of the shared runqueue. The queuing time was shorter than a millisecond, which made direct measurements difficult. We therefore implemented a program that repeats task-popping from a shared queue 268,435,456 times to calculate the average queuing time. Figure 11 shows the result. The queuing time increased as the number of cores increased. That is, a popping request was cancelled when several computation units simultaneously tried to obtain a task from the queue, thanks to the protection technique proposed by Arora et al. [56]. As seen in Fig. 11, the queuing required several micro-seconds whenever a queuing rush occurred.

As seen in Fig. 10, a single core could handle 0.9 observation items per microsecond, which means that the queuing time has a great impact on the lack of acceleration. For the FIFO case, the shared runqueue was accessed frequently by all computation units. For the ADCA, such rushes are rare. This is the reason for the superior robustness against fine-grained parallelization.

4.3   Robustness Against Load Imbalance

To demonstrate the robustness against load imbalance of ADCA, we evaluated strong and weak scaling using the EM algorithm on an HPMM. For the strong scaling, we tested two datasets: `equal` and `slope`. In the `equal` dataset, each segment had the same number of observation items. In the `slope` dataset, the number of observation items in each segment was made proportional to the segment ID, assigned continuously from 1 to either 1024 or 16,384. The datasets comprised 134,217,728 observation items. The grain size was set to 256. For weak scaling, we tested two series of datasets: `fixed` and `scale`. In the `fixed` datasets, the number of segments was constant, regardless of the number of cores. In the `scale` datasets, the number of segments was proportional to the number of cores. The datasets contained 1,048,576 items per core and the grain size was set to 256. Figure 12 shows the results. For both the `equal` and `slope` datasets, ADCA achieved an almost linear speedup. Note that the graphs almost exactly

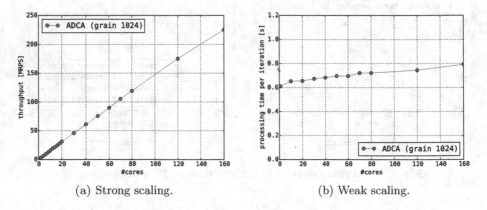

(a) Strong scaling.   (b) Weak scaling.

Fig. 13. Scalability of the EM algorithm on a GMM (`hp160`).

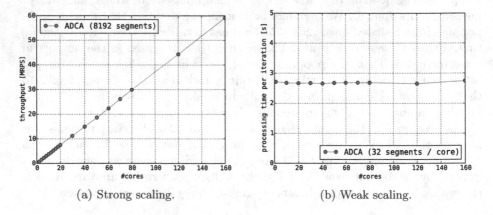

(a) Strong scaling.   (b) Weak scaling.

Fig. 14. Scalability of the EM algorithm on an HPMM (`hp160`).

match each other, with load imbalance having little influence. The weak scaling results demonstrate the great flexibility of ADCA, which is applicable to both *large* tasks handling many observation items and to *small* tasks handling a few observation items. In all cases, the throughput was constant and the processing time was determined only by the size of the dataset.

4.4 Scalability on a 160-Core NUMA Machine

We evaluated strong and weak scaling for our method not only on hu080, but also on hp160. The GMM datasets involved a total of 268,435,456 observation items for strong scaling and 1,048,576 observation items per core for weak scaling. The HPMM datasets involved a total of 134,217,728 observation items for strong scaling and 524,288 observation items per core for weak scaling. In the HPMM case, the number of items in each segment followed a continuous uniform distribution for strong scaling and a Gaussian distribution for weak scaling. The grain size was 1024. Figures 13 and 14 show the results, indicating a near-linear speedup.

5 Conclusions

We have investigated a divide-and-conquer-based parallel computation strategy for machine-learning algorithms. Our approach not only reduces task-scheduling overheads dramatically, but also realizes efficient load balancing by cooperating with a work-stealing scheduler. Furthermore, the divide-and-conquer algorithm derives parameters without requiring mutual exclusion or atomic operations with shared variables by using a buffering solution that avoids the bottleneck of cache-coherence protocols in NUMA environments. We tested the scalability of our approach with both 80-core and 160-core NUMA computers and found that the divide-and-conquer solution achieved far superior scalability to FIFO-based parallelization and showed robustness against load-imbalanced datasets.

In this work, we have evaluated our method only for shared-memory computers with little discussion of reference locality because the buffering solution dealt with much of the reference-locality problem effectively. However, we intend to investigate ADCA for distributed-memory environments in future work, and the buffering solution alone would not be sufficient provision against the greater latency of message passing. Considering the memory access patterns of machine-learning algorithms, there is room for improved reference locality, given that the algorithms access observation items one by one continuously at each learning step, and repeat the steps many times. That is, after the scheduler assigns tasks and observation subsets to nodes before processing, enabling each task to access only its local data, the scheduler could improve reference locality by sending the same tasks to the same nodes at every step, as proposed by Yang et al. [26].

In other future work, our approach will seek to exploit the characteristics of GPUs. As stated in Sect. 2.1, GPUs are hardly applicable to graphical models on their own. Fortunately, a CPU can cooperate with a GPU by using CUDA

42   T. Kawakatsu et al.

[57,58], and a GPU could realize load balancing by cooperating with ADCA through CUDA. As shown in Algorithm 2, our approach employs a loop at the grain level. We expect GPUs to be able to accelerate this loop.

Acknowledgment. This work was supported by the CPS-IIP (http://www.cps.nii. ac.jp.) project under the research promotion program for national challenges *Research and development for the realization of the next-generation IT platforms* of the Ministry of Education, Culture, Sports, Science and Technology (MEXT), Japan. The experimental environment was made available by Assistant Prof. Hajime Imura at the Meme Media Laboratory, Hokkaido University, and Yasuhiro Shirai at HP Japan Inc.

A   General EM Algorithm

A.1   EM on GMM

The GMM is a popular probabilistic model described by a weighted linear sum of K normal distributions:

$$p(\boldsymbol{x}) = \sum_{k=1}^{K} w_k \mathcal{N}(\boldsymbol{x}; \boldsymbol{\mu}_k, S_k), \tag{1}$$

where w_k is the weight, $\boldsymbol{\mu}_k$ is the mean, and S_k is the covariance matrix of the kth normal distribution. An observation item $\boldsymbol{x}$ is generated by a normal distribution selected with a probability of w_k. We transcribe parameters $\theta_k = (w_k, \boldsymbol{\mu}_k, S_k)$ for the sake of simplicity, and θ is the set of all θ_k. The likelihood function $\mathcal{L}(\theta)$ indicates how likely it is that the probabilistic model regenerates the training dataset. Assuming independence among observation items, $\mathcal{L}(\theta)$ is equal to the joint probability of all observation data. $\mathcal{L}$ is defined in log-likelihood terms because $p(\boldsymbol{x}_n|\theta)$ is very small:

$$\mathcal{L}(\theta) = \sum_{n}^{N} \log \sum_{k}^{K} w_k \mathcal{N}(\boldsymbol{x}_n; \boldsymbol{\mu}_k, S_k). \tag{2}$$

In the EM context, we need only maximize $\mathcal{L}$. However, because a GMM is a latent-variable model, it requires step-by-step improvement. The posterior probability q_{nk} that the nth observation item $\boldsymbol{x}_n$ is generated by the kth normal distribution is:

$$q_{nk} = \frac{w_k \mathcal{N}(\boldsymbol{x}_n; \boldsymbol{\mu}_k, S_k)}{\sum_{k}^{K} w_k \mathcal{N}(\boldsymbol{x}_n; \boldsymbol{\mu}_k, S_k)}. \tag{3}$$

Of the two repeated steps, the E-step calculates q_{nk} for all pairs of data $\boldsymbol{x}_n$ and the kth normal distribution, and the M-step updates the parameters as follows:

$$\hat{w}_k = \frac{1}{N} \sum_{n}^{N} q_{nk}, \tag{4}$$

$$\hat{\boldsymbol{\mu}}_k = \frac{1}{N\hat{w}_k} \sum_n^N q_{nk} \boldsymbol{x}_n, \tag{5}$$

$$\hat{S}_k = \frac{1}{N\hat{w}_k} \sum_n^N q_{nk}(\boldsymbol{x}_n - \hat{\boldsymbol{\mu}}_k)^T(\boldsymbol{x}_n - \hat{\boldsymbol{\mu}}_k). \tag{6}$$

The E-step and M-step are repeated alternately until $\mathcal{L}$ converges. In practice, the covariance matrix S_k is assumed to be a diagonal matrix and the calculation is therefore simplified as follows:

$$\hat{S}_{kd} = \frac{1}{N\hat{w}_k} \left(\sum_n^N \hat{q}_{nk} \boldsymbol{x}_{nd}^2 \right) - \hat{\mu}_{kd}^2. \tag{7}$$

In the E-step, $N \times K$ q_{nk} is calculated, and in the M-step, q_{nk} is summed in the N axis and the parameter θ_k is updated. However, the posterior table can be too large and can exceed the hard-disk capacity when N is very large. Because of poor memory throughput, the processing speed will then degrade greatly. To avoid this condition, the parallel EM algorithm requires a large memory space.

A.2 EM on HPMM

Kinoshita et al. used an HPMM to detect traffic incidents [10]. They assumed that probe-car records follow a hierarchical PMM and that each road segment has its own local parameters. In their model, the probability of a single record $\boldsymbol{x}$ in a segment s is described as follows:

$$p(\boldsymbol{x}|s) = \sum_{k=1}^K w_{sk}\mathcal{P}(\boldsymbol{x}; \boldsymbol{\mu}_k), \tag{8}$$

where w_{sk} is the kth Poisson distribution's weight in segment s, and $\boldsymbol{\mu}_k$ is the kth Poisson distribution's mean. w_{sk} is particular to the segment, whereas $\boldsymbol{\mu}_k$ is common to all segments. The log-likelihood $\mathcal{L}(\theta)$ is defined as follows:

$$\mathcal{L}(\theta) = \sum_{s=1}^S \sum_{n=1}^{N_s} \log \sum_{k=1}^K w_{sk}\mathcal{P}(\boldsymbol{x}_{sn}; \boldsymbol{\mu}_k), \tag{9}$$

where N_s is the number of records in segment s. As for GMMs, we must calculate the posterior probability q_{snk} that the nth record $\boldsymbol{x}_{sn}$ in segment s is generated by the kth Poisson distribution for all pairs of (s, n, k) in each E-step:

$$q_{snk} = \frac{w_{sk}\mathcal{P}(\boldsymbol{x}_{sn}; \boldsymbol{\mu}_k)}{\displaystyle\sum_{k=1}^K w_{sk}\mathcal{P}(\boldsymbol{x}_{sn}; \boldsymbol{\mu}_k)}. \tag{10}$$

In the M-step, the weight w_{sk} and mean $\boldsymbol{\mu}_k$ are recalculated:

$$\hat{w}_{sk} = \frac{1}{N_s} \sum_{n=1}^{N_s} q_{snk}, \tag{11}$$

$$\hat{\boldsymbol{\mu}}_k = \frac{\displaystyle\sum_{s=1}^{S} \sum_{n=1}^{N_s} q_{snk}\boldsymbol{x}_{sn}}{\displaystyle\sum_{s=1}^{S} \sum_{n=1}^{N_s} q_{snk}}. \tag{12}$$

Each road segment has a massive number of records, with the actual number varying greatly from segment to segment. This implies that we should take measures against load imbalance.

References

1. Zaharia, M., Chowdhury, M., Franklin, M.J., Shenkerand, S., Stoica, I.: Spark: cluster computing with working sets. In: Proceedings of the 2nd USENIX Conference on Hot Topics in Cloud Computing, June 2010
2. Zaharia, M., Chowdhury, M., Das, T., Dave, A., Ma, J., MacCauley, M., Franklin, M.J., Shenker, S., Stoica, I.: Resilient distributed datasets: a fault-tolerant abstraction for in-memory cluster computing. In: Proceedings of the 9th USENIX Conference on Networked Systems Design and Implementation, April 2012
3. Power, R., Li, J.: Piccolo: building fast, distributed programs with partitioned tables. In: Proceedings of the 9th USENIX Conference on Operating Systems Design and Implementation, October 2010
4. Huang, C., Chen, Q., Wang, Z., Power, R., Ortiz, J., Li, J., Xiao, Z.: Spartan: a distributed array framework with smart tiling. In: Proceedings of the USENIX Annual Technical Conference, July 2015
5. Dijkstra, E.W.: Cooperating sequential processes. EWD: EWD123 (1968)
6. Mohr, E., Kranz Jr., D.A., Halstead, R.H.: Lazy task creation: a technique for increasing the granularity of parallel programs. In: Proceedings of the 1990 ACM Conference on LISP and Functional Programming, May 1990
7. Blumofe, R.D., Joerg, C.F., Kuszmaul, B.C., Leiserson, C.E., Randall, K.H., Zhou, Y.: Cilk: an efficient multithreaded runtime system. In: Proceedings of the Fifth ACM SIGPLAN Symposium on Principles and Practice of Parallel Programming, August 1995
8. Dempster, A.P., Laird, N.M., Rubin, D.B.: Maximum likelihood from incomplete data via the EM algorithm. J. Roy. Stat. Soc. Ser. B (Methodol.) **39**(1), 1–38 (1977)
9. McLachlan, G.J., Krishnan, T.: The EM Algorithm and Extensions. Wiley, Hoboken (2008)
10. Kinoshita, A., Takasu, A., Adachi, J.: Traffic incident detection using probabilistic topic model. In: Proceedings of the Workshops of the EDBT/ICDT 2014 Joint Conference, March 2014

11. Kinoshita, A., Takasu, A., Adachi, J.: Real-time traffic incident detection using a probabilistic topic model. Inf. Syst. **54**(C), 169–188 (2015)
12. Pereira, S.S., Lopez-Valcarce, R., Pages-Zamora, A.: A diffusion-based EM algorithm for distributed estimation in unreliable sensor networks. IEEE Signal Process. Lett. **20**(6), 595–598 (2013)
13. Chen, J., Salim, M.B., Matsumoto, M.: A gaussian mixture model-based continuous boundary detection for 3d sensor networks. Sensors **10**(8), 7632–7650 (2010)
14. Miura, K., Noguchi, H., Kawaguchi, H., Yoshimoto, M.: A low memory bandwidth gaussian mixture model (GMM) processor for 20,000-word real-time speech recognition FPGA system. In: 2008 International Conference on ICECE Technology, December 2008
15. Gupta, K., Owens, J.D.: Three-layer optimizations for fast GMM computations on GPU-like parallel processors. In: IEEE Workshop on Automatic Speech Recognition & Understanding, December 2009
16. Stauffer, C., Grimson, W.E.L.: Adaptive background mixture models for real-time tracking. In: IEEE Computer Society Conference on Computer Vision and Pattern Recognition, June 1999
17. Li, H., Achim, A., Bull, D.R.: GMM-based efficient foreground detection with adaptive region update. In: Proceedings of the 16th IEEE International Conference on Image Processing, November 2009
18. Patel, C.I., Patel, R.: Gaussian mixture model based moving object detection from video sequence. In: Proceedings of the International Conference and Workshop on Emerging Trends in Technology, February 2011
19. Song, Y., Li, X., Liu, Q.: Fast moving object detection using improved gaussian mixture models. In: International Conference on Audio, Language and Image Processing, July 2014
20. Rumelhart, D.E., Hinton, G.E., Williams, R.J.: Learning representations by backpropagating errors. In: Neurocomputing: Foundations of Research, January 1988
21. Liu, Z., Li, H., Miao, G.: MapReduce-based backpropagation neural network over large scale mobile data. In: Sixth International Conference on Natural Computation, August 2010
22. Gu, R., Shen, F., Huang, Y.: A parallel computing platform for training large scale neural networks. In: IEEE International Conference on Big Data, October 2013
23. Hillis, W.D., Steele Jr., G.L.: Data parallel algorithms. Commun. ACM Spec. Issue Parallelism **29**(12), 1170–1183 (1986)
24. Flynn, M.J.: Some computer organizations and their effectiveness. IEEE Trans. Comput. **C–21**(9), 948–960 (1972)
25. Kwedlo, W.: A parallel EM algorithm for Gaussian mixture models implemented on a NUMA system using OpenMP. In: 22nd Euromicro International Conference on Parallel, Distributed and Network-Based Processing (PDP), February 2014
26. Yang, R., Xiong, T., Chen, T., Huang, Z., Feng, S.: DISTRIM: parallel GMM learning on multicore cluster. In: IEEE International Conference on Computer Science and Automation Engineering (CSAE), May 2012
27. Wolfe, J., Haghighi, A., Klein, D.: Fully distributed EM for very large datasets. In: Proceedings of the 25th International Conference on Machine Learning, July 2008
28. Kumar, N.S.L.P., Satoor, S., Buck, L.: Fast parallel expectation maximization for gaussian mixture models on GPUs using CUDA. In: 11th IEEE International Conference on High Performance Computing and Communications, June 2009

29. Machlica, L., Vanek, J., Zajic, Z.: Fast estimation of gaussian mixture model parameters on GPU using CUDA. In: 12th International Conference on Parallel and Distributed Computing, Applications and Technologies (PDCAT), October 2011
30. Altinigneli, M.C., Plant, C., Bohm, C.: Massively parallel expectation maximization using graphics processing units. In: Proceedings of the 19th ACM SIGKDD International Conference on Knowledge Discovery and Data Mining, August 2013
31. Bergstrom, L., Reppy, J.: Nested data-parallelism on the GPU. In: Proceedings of the 17th ACM SIGPLAN International Conference on Functional Programming, September 2012
32. Lee, H., Brown, K.J., Sujeeth, A.K., Rompf, T., Olkotun, K.: Locality-aware mapping of nested parallel patterns on GPU. In: Proceedings of eht 47th Annual IEEE/ACM International Symposium on Microarchitecture, December 2014
33. Feeley, M.: A message passing implementation of lazy task creation. In: Halstead, R.H., Ito, T. (eds.) PSC 1992. LNCS, vol. 748, pp. 94–107. Springer, Heidelberg (1993). doi:10.1007/BFb0018649
34. Umatani, S., Yasugi, M., Komiya, T., Yuasa, T.: Pursuing laziness for efficient implementation of modern multithreaded languages. In: Veidenbaum, A., Joe, K., Amano, H., Aiso, H. (eds.) ISHPC 2003. LNCS, vol. 2858, pp. 174–188. Springer, Heidelberg (2003). doi:10.1007/978-3-540-39707-6_13
35. Acar, U.A., Chargueraud, A., Rainey, M.: Scheduling parallel programs by work stealing with private deques. In: Proceedings of the 18th ACM SIGPLAN Symposium on Principles and Practice of Parallel Programming, February 2013
36. Frigo, M., Leiserson, C.E., Randall, K.H.: The implementation of the Cilk-5 multithreaded language. In: Proceedings of the ACM SIGPLAN 1998 Conference on Programming Language Design and Implementation, May 1998
37. Min, S.J., Iancu, C., Yelick, K.: Hierarchical work stealing on manycore clusters. In: Fifth Conference on Partitioned Global Address Space Programming Models, October 2011
38. Olivier, S.L., Porterfield, A.K., Wheeler, K.B., Prins, J.F.: Scheduling task parallelism on multi-socket multicore systems. In: Proceedings of the 1st International Workshop on Runtime and Operating Systems for Supercomputers, May 2011
39. Olivier, S.L., Porterfield, A.K., Wheeler, K.B., Spiegel, M., Prins, J.F.: OpenMP task scheduling strategies for multicore numa systems. Int. J. High Perform. Comput. Appl. **26**(2), 110–124 (2012)
40. Nakashima, J., Nakatani, S., Taura, K.: Design and implementation of a customizable work stealing scheduler. In: 3rd International Workshop on Runtime and Operating Systems for Supercomputers, June 2013
41. Kranz, D.A., Halstead, R.H., Mohr Jr., E.: Mul-T: a high-performance parallel lisp. In: Proceedings of the ACM SIGPLAN 1989 Conference on Programming Language Design and Implementation, June 1989
42. Wheeler, K.B., Murphy, R.C., Thain, D.: Qthreads: an API for programming with millions of lightweight threads. In: IEEE International Symposium on Parallel and Distributed Processing, April 2008
43. Molka, D., Hackenberg, D., Shone, R., Muller, M.S.: Memory performance and cache coherency effects on an intel nahalem multiprocessor system. In: 18th International Conference on Parallel Architectures and Compilation Techniques, September 2009
44. Molka, D., Hackenberg, D., Schone, R., Nagel, W.E.: Cache coherence protocol and memory performance of the intel haswell-EP architecture. In: 44th International Conference on Parallel Processing, September 2015

45. Charles, P., Donawa, C., Ebcioglu, K., Grothoff, C., Kielstra, A., von Praun, C., Saraswat, V., Sarkar, V.: X10: an object-oriented approach to non-uniform cluster computing. In: Proceedings of the 20th Annual ACM SIGPLAN Conference on Object-Oriented Programming, Systems, Languages, and Applications, October 2005

46. Callahan, D., Chamberlain, B.L., Zima, H.P.: The cascade high productivity language. In: 9th International Workshop on High-Level Parallel Programming Models and Supportive Environments, April 2004

47. Dean, J., Ghemawat, S.: MapReduce: simplified data processing on large clusters. In: Proceedings of the 6th Conference on Symposium on Opearting Systems Design & Implementation, vol. 6, December 2004

48. Furmento, N., Goglin, B.: Enabling high-performance memory migration for multithreaded applications on Linux. In: IEEE International Symposium on Parallel & Distributed Processing, May 2009

49. Lameter, C.: NUMA (non-uniform memory access): an overview. Queue **11**(7), 40 (2013)

50. Low, Y., Gonzalez, J., Kyrola, A., Bickson, D., Guestrin, C., Hellerstein, J.: GraphLab: a new framework for parallel machine learning. In: Proceedings of the 26th Conference on Uncertainty in Artificial Intelligence, June 2010

51. Low, Y., Bickson, D., Gonzalez, J., Guestrin, C., Kyrola, A., Hellerstein, J.M.: Distributed GraphLab: a framework for machine learning and data mining in the cloud. In: Proceedings of the VLDB Endowment, April 2012

52. Hamidouche, K., Falcou, J., Etiemble, D.: A framework for an automatic hybrid MPI+ openMP code generation. In: Proceedings of the 19th High Performance Computing Symposia, April 2011

53. Si, M., Pena, A.J., Balaji, P., Takagi, M., Ishikawa, Y.: MT-MPI: multithreaded MPI for many-core environments. In: Proceedings of the 28th ACM International Conference on Supercomputing, June 2014

54. Luo, M., Lu, X., Hamidouche, K., Kandalla, K., Panda, D.K.: Initial study of multi-endpoint runtime for MPI+ openMP hybrid programming model on multi-core systems. In: Proceedings of the 19th ACM SIGPLAN Symposium on Principles and Practice of Parallel Programming, February 2014

55. Kawakatsu, T., Kinoshita, A., Takasu, A., Adachi, J.: Highly efficient parallel framework: a divide-and-conquer approach. In: Chen, Q., Hameurlain, A., Toumani, F., Wagner, R., Decker, H. (eds.) DEXA 2015. LNCS, vol. 9262, pp. 162–176. Springer, Heidelberg (2015). doi:10.1007/978-3-319-22852-5_15

56. Arora, N.S., Blumofe, R.D., Plaxton, C.G.: Thread scheduling for multiprogrammed multiprocessors. In: Proceedings of the Tenth Annual ACM Symposium on Parallel Algorithms and Architectures, June 1998

57. Kirk, D.B., Hwu, W.W.: Processors, Programming Massively Parallel: A Hands-on Approach. Morgan Kaufmann, San Francisco (2010)

58. Nvidia. CUDA C programming guide version 6.5, August 2014

Multistore Big Data Integration
with CloudMdsQL

Carlyna Bondiombouy, Boyan Kolev$^{(\boxtimes)}$, Oleksandra Levchenko,
and Patrick Valduriez

Inria and LIRMM, University of Montpellier, Montpellier, France
{carlyna.bondiombouy,boyan.kolev,
oleksandra.levchenko,patrick.valduriez}@inria.fr

Abstract. Multistore systems have been recently proposed to provide integrated access to multiple, heterogeneous data stores through a single query engine. In particular, much attention is being paid on the integration of unstructured big data typically stored in HDFS with relational data. One main solution is to use a relational query engine that allows SQL-like queries to retrieve data from HDFS, which requires the system to provide a relational view of the unstructured data and hence is not always feasible. In this paper, we propose a functional SQL-like query language (based on CloudMdsQL) that can integrate data retrieved from different data stores, to take full advantage of the functionality of the underlying data processing frameworks by allowing the ad-hoc usage of user defined map/filter/reduce operators in combination with traditional SQL statements. Furthermore, our solution allows for optimization by enabling subquery rewriting so that bind join can be used and filter conditions can be pushed down and applied by the data processing framework as early as possible. We validate our approach through implementation and experimental validation with three data stores and representative queries. The experimental results demonstrate the usability of the query language and the benefits from query optimization.

1 Introduction

A major trend in cloud computing and big data is the understanding that there is "no one size fits all" solution. Thus, there has been a blooming of different cloud data management solutions, such as NoSQL, distributed file systems (e.g. Hadoop HDFS), and big data processing frameworks (e.g. Hadoop MapReduce or Apache Spark), specialized for different kinds of data and able to perform orders of magnitude better than traditional RDBMS. However, this has led to a wide diversification of data store interfaces and the loss of a common programming paradigm. This makes it very hard for a user to integrate and analyze her data sitting in different data stores, e.g. RDBMS, NoSQL and HDFS. To address this problem, multistore systems [1, 8, 9, 11–15] have been recently proposed to provide integrated access to multiple, heterogeneous data stores through a single query engine.

Compared to multidatabase systems [16], multistore systems typically trade source autonomy for efficiency, using a tightly-coupled approach. In particular, much attention is being paid on the integration of unstructured big data (e.g. produced by web

© Springer-Verlag Berlin Heidelberg 2016
A. Hameurlain et al. (Eds.): TLDKS XXVIII, LNCS 9940, pp. 48–74, 2016.
DOI: 10.1007/978-3-662-53455-7_3

applications) typically stored in HDFS with relational data, e.g. in a data warehouse. One main solution is to use a relational query engine (e.g. Apache Hive) on top of a data processing framework (e.g. Hadoop MapReduce), which allows SQL-like queries to retrieve data from HDFS. However, this requires the system to provide a relational view of the unstructured data, which is not always feasible. In case the data store is managed independently from the relational query processing system, complex data transformations may need to take place (e.g. by applying specific map-reduce jobs) before the data can be processed by means of relational operators. Let us illustrate the problem, which will be the focus of this paper, with the following scenario.

Example scenario. An editorial office needs to find appropriate reporters for a list of publications based on given keywords. For the purpose, the editors need an analysis of the logs from a scientific forum stored in a Hadoop cluster in the cloud to find experts in a certain research field, considering the users who have mentioned particular keywords most frequently; and these results must be joined to the relational data in an RDBMS containing author and publication information. However, the forum application keeps log data about its posts in a non-tabular structure (the left side of the example below), namely in text files where a single record corresponds to one post and contains a fixed number of fields about the post itself (timestamp and username in the example) followed by a variable number of fields storing the keywords mentioned in the post.

```
                                               KW        expert   freq
2014-12-13, alice, storage, cloud              cloud     alice    2
2014-12-22, bob, cloud, virtual, app   ─────▶  storage   alice    1
2014-12-24, alice, cloud                       virtual   bob      1
                                               app       bob      1
```

The unstructured log data needs to be transformed into a tabular dataset containing for each keyword the expert who mentioned it most frequently (the right side of the example above). Such transformation requires the use of programming techniques like chaining map/reduce operations that should take place before the data is involved in relational operators. Then the result dataset will be ready to be joined with the publication data retrieved from the RDBMS in order to suggest an appropriate reviewer for each publication. Being able to request such data processing with a single query is the scenario that motivates our work. However, the challenge in front of the query processor is optimization, i.e. it should be able of analyzing the operator execution flow of a query and performing operation reordering to take advantage of well-known optimization techniques (e.g. selection pushdowns and use of semi-joins) in order to yield efficient query execution.

Existing solutions to integrate such unstructured and structured data do not directly apply to solve our problem, as they rely on having a relational view of the unstructured data, and hence require complex transformations. SQL engines, such as Hive, on top of distributed data processing frameworks are not always capable of querying unstructured HDFS data, thereby forcing the user to query the data by defining map/reduce functions.

Our approach is different as we propose a query language that can directly express subqueries that can take full advantage of the functionality of the underlying data processing frameworks. Furthermore, the language should allow for query optimization, so that the query operator execution sequence specified by the user may be reordered by taking into account the properties of map/filter/reduce operators together with the properties of relational operators. This is especially useful for applying efficient query optimization by exploiting bind joins [10]; and we pay special attention to this throughout our experimental evaluation. Finally, we want to respect the autonomy of the data stores, e.g. HDFS and RDBMS, so that they can be accessible and controlled from outside our query engine with their own interface.

In this paper, we propose a functional SQL-like query language (based on CloudMdsQL) and query engine to retrieve data from two different kinds of data stores – an RDBMS and a distributed data processing framework such as Apache Spark or Hadoop MapReduce on top of HDFS – and combine them by applying data integration operators (mostly joins). We assume that each data store is fully autonomous, i.e. the query engine has no control over the structure and organization of data in the data stores. For this reason, the architecture of our query engine is based on the traditional mediator/wrapper architectural approach [21] that abstracts the query engine from the specifics of each of the underlying data stores. However, users need to be aware of how data are organized across the data stores, so that they write valid queries. A single query of our language can request data to be retrieved from both stores and then a join to be performed over the retrieved datasets. The query therefore contains embedded invocations to the underlying data stores, expressed as subqueries. As our query language is functional, it introduces a tight coupling between data and functions. A subquery, addressing the data processing framework, is represented by a sequence of map/filter/ reduce operations, expressed in a formal notation. On the other hand, SQL is used to express subqueries that address the relational data store as well as the main statement that performs the integration of data retrieved by all subqueries. Thus, a query benefits from both high expressivity (by allowing the ad-hoc usage of user defined map/ filter/reduce operators in combination with traditional SQL statements) and optimizability (by enabling subquery rewriting so that bind join and filter conditions can be pushed inside and executed at the data store as early as possible).

This paper is a major extension of [4], with an improved generic architecture of the query engine (to support a wider range of underlying data models and to provide a tighter coupling with the data processing framework), a real experimental validation with three data stores (relational, document, and HDFS) and queries across them, and a more detailed comparison with the state of the art.

The rest of this paper is organized as follows. Section 2 introduces the language and its notation to express map/filter/reduce subqueries. Section 3 presents the architecture of the query engine. Section 4 elaborates more on the query processing and presents the properties of map/filter/reduce operators that constitute rewrite rules to perform query optimization. Section 5 gives a use case example walkthrough. Section 6 presents an experimental validation with (semi-)structured data stored in PostgreSQL and MongoDB, and unstructured data stored in an HDFS cluster and processed using Apache Spark. Section 7 discusses related work. Section 8 concludes.

2 Query Language

The query language is based on a more general common query language, called CloudMdsQL [12], designed in the context of the CoherentPaaS project [7] to solve the problem of querying multiple heterogeneous databases (e.g. relational and NoSQL) within a single query while preserving the expressivity of their local query mechanisms. The common language itself is SQL-based with the extended capabilities for embedding subqueries expressed in terms of each data store's native query interface. The common data model respectively is table-based, with support of rich datatypes that can capture a wide range of the underlying data stores' datatypes, such as MongoDB arrays and JSON objects, in order to handle non-flat and nested data, with basic operators over such composite datatypes.

In this section, we introduce a formal notation to define Map/Filter/Reduce (MFR) subqueries in CloudMdsQL that request data processing in an underlying big data processing framework (DPF). Then we give an overview of how MFR statements are combined with SQL statements to express integration queries against a relational database and a DPF. Notice that the data processing defined in an MFR statement is not executed by the query engine, but is meant to be translated to a sequence of invocations to API functions of the DPF. In this paper, we use Apache Spark as an example of DPF, but the concept can be generalized to a wider range of frameworks that support the MapReduce programming model (such as Hadoop MapReduce, CouchDB, etc.).

2.1 MFR Notation

An MFR statement represents a sequence of MFR operations on datasets. A dataset is considered simply as an abstraction for a set of tuples, where a tuple is a list of values, each of which can be a scalar value or another tuple. Although tuples can generally have any number of elements, mostly datasets that consist of key-value tuples are being processed by MFR operations. In terms of Apache Spark, a dataset corresponds to an RDD (Resilient Distributed Dataset – the basic programming unit of Spark). Each of the three major MFR operations (MAP, FILTER and REDUCE) takes as input a dataset and produces another dataset by performing the corresponding transformation. Therefore, for each operation there should be specified the transformation that needs to be applied on tuples from the input dataset to produce the output tuples. Normally, a transformation is expressed with an SQL-like expression that involves special variables; however, more specific transformations may be defined through the use of lambda functions.

Core operators. The MAP operator produces key-value tuples by performing a specified transformation on the input tuples. The transformation is defined as an SQL-like expression that will be evaluated for each tuple of the input data set and should return a pair of values. The special variable TUPLE refers to the input tuple and its elements are addressed using a bracket notation. Moreover, the variables KEY and VALUE may be used as aliases to TUPLE[0] and TUPLE[1] respectively. The FILTER operator selects from the input tuples only those, for which a specified condition is evaluated to *true*. The filter condition is defined as a boolean expression using the same special variables TUPLE, KEY, and VALUE. The REDUCE operator performs aggregation on values associated with the same key and produces a key-value dataset where each key is unique. The reduce transformation may be specified as an aggregate function (SUM, AVG, MIN, MAX or COUNT). Similarly to MAP, two other mapping operators are introduced: FLAT_MAP may produce numerous output tuples for a single input tuple; and MAP_VALUES defines a transformation that preserves the keys, i.e. applicable only to the values.

Let us consider the following simple example inspired by the popular MapReduce tutorial application "word count". We assume that the input dataset for the MFR statement is a list of words. To count the words that contain the string 'cloud', we write the following composition of MFR operations:

```
MAP(KEY, 1).FILTER( KEY LIKE '%cloud%' ).REDUCE( SUM )
```

The first operation transforms each tuple (which has a single word as its only element) of the input dataset into a key-value pair where the word is mapped to a value of 1. The second operation selects only those key-value pairs for which the key contains the string 'cloud'. And the third one groups all tuples by key and performs a sum aggregate on the values for each key.

To process this statement, the query engine first looks for opportunities to optimize the execution by operator reordering. By applying MFR rewrite rules (explained in detail in Sect. 4.2), it finds out that the FILTER and MAP operations may be swapped so that the filtering is applied at an earlier stage. Further, it translates the sequence of operations into invocations of the underlying DPF's API. Notice that whenever a REDUCE transformation function has the associative property (like the SUM function), an additional combiner function call may be generated that precedes the actual reducer, so that as much data as possible will be reduced locally; e.g., this would be valid in the case of Hadoop MapReduce as the DPF, because it does not automatically perform local reduce. In the case of Apache Spark as the DPF, the query engine generates the following Python fragment to be included in a script that will be executed in Spark's Python environment:

```
dataset.filter( lambda k: 'cloud' in k ) \
       .map( lambda k: (k, 1) ) \
       .reduceByKey( lambda a, b: a + b )
```

In this example, all the MFR operations are translated to their corresponding Spark functions and all transformation expressions are translated to Python anonymous functions. In fact, to increase its expressivity, the MFR notation allows direct usage of anonymous functions to specify transformation expressions. This allows user-defined mapping functions, filter predicates, or aggregates to be used in an MFR statement. The user, however, needs to be aware of how the query engine is configured to interface the DPF, in order to know which language to use for the definition of inline anonymous functions (e.g. Spark may be used with Python or Scala, CouchDB – with JavaScript, etc.).

Input/output operators are normally used for transformation of data before and after the core map/filter/reduce execution chain. The SCAN operator loads data from its storage and transforms it to a dataset ready to be consumed by a core MFR operator. The PROJECT operator converts a key-value dataset to a tabular dataset ready to be involved in relational operations.

2.2   Combining SQL and MFR

Queries that integrate data from both a relational data store and a DPF usually consist of two subqueries (one expressed in SQL that addresses the relational database and another expressed in MFR that addresses the DPF) and an integration SELECT statement. The syntax follows the CloudMdsQL grammar introduced in [12]. A sub-query is defined as a named table expression, i.e. an expression that returns a table and has a name and signature. The signature defines the names and types of the columns of the returned relation. Thus, each query, although agnostic to the underlying data stores' schemas, is executed in the context of an ad-hoc schema, formed by all named table expressions within the query. A named table expression can be defined by means of either an SQL SELECT statement (that the query compiler is able to analyze and possibly rewrite) or a native expression (that the query compiler considers as a black box and passes to the wrapper as is, thus delegating it the processing of the subquery).

In this paper, we extend the usability of CloudMdsQL by adding the capability of handling MFR subqueries against DPFs and combining them with subqueries against other data stores. This is done in full compliance with CloudMdsQL properties, such as the ability to express nested subqueries (so that the output of one subquery, e.g. against an RDBMS, can be used as input to another subquery, e.g. MFR) which we further illustrate by the usage of bind joins. MFR subqueries are expressed as native named table expressions; this means that they are passed to their corresponding wrappers to process them (explained in more detail in Sect. 3).

In general, a single query can address a number of data stores by containing several named table expressions. We will now illustrate with a simple example how SQL and MFR statements can be combined, and in Sect. 5 will focus on a more sophisticated example involving 3 data stores. The following sample query contains two subqueries, defined by the named table expressions T1 and T2, and addressed respectively against the data stores aliased with identifiers rdb (for the SQL database) and hdfs (for the DPF):

```
T1(title string, kw string)@rdb = ( SELECT title, kw FROM tbl )
T2(word string, count int)@hdfs = {*
   SCAN(TEXT,'words.txt')
      .MAP(KEY,1).REDUCE(SUM).PROJECT(KEY,VALUE)
*}
SELECT title, kw, count FROM T1 JOIN T2 ON T1.kw = T2.word
WHERE T1.kw LIKE '%cloud%'
```

The purpose of this query is to perform relational algebra operations (expressed in the main SELECT statement) on two datasets retrieved from a relational database and a DPF. The two subqueries are sent independently for execution against their data stores in order the retrieved relations to be joined by the query engine. The SQL table expression T1 is defined by an SQL subquery. T2 is an MFR expression that requests data retrieval from a text source and data processing by the specified map/reduce operations. Both subqueries are subject to rewriting by pushing into it the filter condition kw LIKE '%cloud%', specified in the main SELECT statement, thus reducing the amount of the retrieved data by increasing the subquery selectivity and the overall efficiency. The so retrieved datasets are then converted to relations following their corresponding signatures, so that the main SELECT statement can be processed with semantic correctness. The PROJECT operator in the MFR statement provides a mapping between the dataset fields and the named table expression columns.

3 Generic Query Engine Architecture

The dominant state-of-the-art architectural model that addresses the problem of data integration and query processing across a diverse set of data stores is the mediator/wrapper architecture. A mediator is a software module that exploits encoded knowledge about certain sets or subsets of data to create information for a higher layer of applications [16]. In addition, a wrapper or adapter is a software component that encapsulates and hides the underlying complexity of sets or subsets of data by means of well-defined interfaces (it establishes communication and a data flow between mediators and data stores). In this section, we briefly describe the generic architecture of our system with an overview of the required steps to process a query.

The query language presented hereby assumes a query engine that follows the traditional mediator/wrapper architectural approach. By explicitly naming a data store identifier in a named table expression's signature, the query addresses the specific wrapper that is preliminarily configured and responsible for handling subqueries against the corresponding data store. Thus, a query can express an integration of data across several data stores, and in particular, integration of structured (relational DB), semi-structured (document DB), and unstructured (distributed storage, based on HDFS) data, which is the case that we focus on throughout our experimental validation.

Figure 1 depicts the corresponding system architecture, containing a CloudMdsQL compiler, a common query processor (the mediator), three wrappers, and the three data stores – a distributed data processing framework (DPF), an RDBMS, and a document data store. The DPF is in charge of performing parallel data processing over a distributed data store. In this architecture, each data source has an associated wrapper that is responsible for executing subqueries against the data store and converting the retrieved datasets to tables matching the requested number and types of columns, so that they are ready to be consumed by relational operators at the query processor. The query processor consumes the query execution plan generated by the compiler and interacts with the wrappers through a common interface to: request handling of subqueries, centralize the information provided by the wrappers, and integrate the subqueries' results. The wrappers transform subqueries provided via the common interface into queries for the data stores. This generic architecture gives us the possibility to use a specific implementation of the query processor and DPF wrapper, while reusing the CloudMdsQL query compiler and wrappers for relational and document data stores

[12]. Although we can also reuse the CloudMdsQL query engine that has a distributed architecture [12], in our experimental work we explore the possibility to adapt the parallel SQL engine Spark SQL [2] to serve as the query processor, thus providing a tighter coupling between the query processor and the underlying DPF and hence taking more advantage of massive parallelism when joining HDFS with relational and document data.

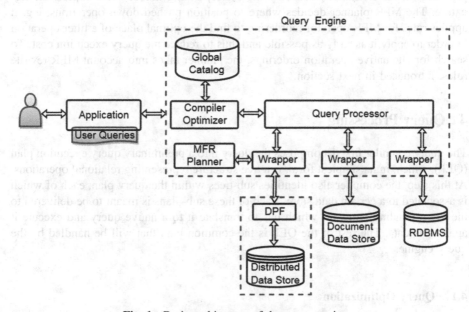

Fig. 1. Basic architecture of the query engine

Each of the wrappers is responsible for completing the execution of subqueries and retrieving the results. Upon initialization, each wrapper may provide to the query compiler the capability of its data store to process pushed down operations [12]. In our setup, all the three wrappers can accept pushdowns of filter predicates. Both the relational and document data store wrappers accept requests from the query processor in the form of query execution sub-plans represented as trees of relational algebra operators, resulting from the compilation of the SELECT statements expressed in the corresponding SQL named table expressions. The sub-plans may include selection operations resulting from pushed down predicates. The wrapper of the relational database has to build a SELECT statement out of a query sub-plan and to run it against its data store; then it retrieves the datasets and delivers them to the query processor in the corresponding format. The wrapper of the document data store (in our case, MongoDB) has to translate the sequence of relational operators from a query sub-plan to the corresponding sequence of MongoDB API calls; then it converts the resulting documents to tuples that match the signature of the corresponding named table expression [12].

The wrapper of the distributed data processing framework has a slightly different behavior as it processes MFR expressions wrapped in native subqueries. First it parses and interprets a subquery written in MFR notation; then uses the MFR planner to find optimization opportunities; and finally translates the resulting sequence of MFR operations to a sequence of DPF's API methods to be executed. Once a dataset is retrieved as a result of the subquery execution, the wrapper provides it to the query processor in the format requested by the corresponding named table expression signature. The MFR planner decides where to position pushed down operations; e.g. it applies rules for MFR operator reordering to find the optimal place of a filter operation in order to apply it as early as possible and thus to reduce the query execution cost. To search for alternative operation orderings, the planner takes into account MFR rewrite rules, introduced in next section.

4   Query Processing

The query compiler first decomposes the query into a preliminary query execution plan (QEP), which, in its simplest form, is a tree structure representing relational operations. At this step, the compiler also identifies sub-trees within the query plan, each of which is associated to a certain data store. Each of these sub-plans is meant to be delivered to the corresponding wrapper, which has to translate it to a native query and execute it against its data. The rest of the QEP is the common plan that will be handled by the query engine.

4.1   Query Optimization

Before its actual execution, a QEP may be rewritten by the query optimizer. To compare alternative rewritings of a query, the optimizer uses a simple catalog, which provides basic information about data store collections such as cardinalities, attribute selectivities and indexes, and a simple cost model. Because of the autonomy of the underlying data stores, in order to derive local cost models, various classical black-box approaches for heterogeneous cost modeling, such as probing [26] and sampling [25, 27], have been adopted by the query optimizer. Thus, cost information can be collected by the wrappers and exposed to the optimizer in the form of cost functions or database statistics. Furthermore, the query language allows for user-defined cost and selectivity functions. And in case of lack of any cost information, heuristic rules are applied.

In our concrete example scenario with PostgreSQL, MongoDB, and MFR subqueries, we use the following strategy. The query optimizer executes an EXPLAIN request to PostgreSQL to directly estimate the cost of a subquery. The MongoDB wrapper runs in background probing queries to collect cardinalities of document collections, index availabilities, and index value distributions (to compute selectivities) and caches them in the query engine's catalog. As for an MFR subquery, if there is no user-provided cost information, the optimizer assumes that it is more expensive than SQL subqueries and plans it at the end of the join order, which would also potentially benefit from the execution of bind joins.

The search space explored for optimization is the set of all possible rewritings of the initial query, by pushing down select operations, expressing bind joins, and join ordering. Unlike in traditional query optimization where many different permutations are possible, this search space is not very large, so we use a simple exhaustive search strategy.

Subquery rewriting takes place in order to request early execution of some operators and thus to increase its overall efficiency. Although several operations are subject to pushdowns across subqueries, in this paper we concentrate on the inclusion of only filter operations inside an MFR subquery. Generally, this is done in two stages: first, the query processor determines which operations can be pushed down for remote execution at the data stores; and second, the MFR planner may further determine the optimal place for inclusion of pushed down operations within the MFR operator chain by applying MFR rewrite rules (explained later in this section). Pushing a selection operation inside a subquery, either in SQL query or MFR operation chain, is usually considered beneficial, because it delegates the selection directly to the data store, which allows for early reducing of the size of data processed and retrieved from the data stores.

4.2 MFR Rewrite Rules

In this section, we introduce and enumerate some rules for reordering of MFR operators, based on their algebraic properties. These rules are used by the MFR planner to optimize an MFR subquery after a selection pushdown takes place.

Rule #1 (name substitution): upon pushdown, the filter is included just before the PROJECT operator and the filter predicate expression is rewritten by substituting column names with references to dataset fields as per the mapping defined by the PROJECT expressions. After this initial inclusion, other rules apply to determine whether it can be moved even farther. Example:

```
T1(a int, b int)@db1 ={* … .PROJECT(KEY, VALUE[0]) *}
SELECT a, b FROM T1 WHERE a > b
```
is rewritten to:

```
T1(a int, b int)@db1 ={* … .FILTER(KEY>VALUE[0]).PROJECT(KEY,VALUE[0]) *}
SELECT a, b FROM T1
```

Rule #2: REDUCE(<transformation>).FILTER(<predicate>) is equivalent to FILTER(<predicate>).REDUCE(<transformation>), if predicate condition is a function only of the KEY, because thus, applying the FILTER before the REDUCE will preserve the values associated to those keys that satisfy the filter condition as they would be if the FILTER was applied after the REDUCE. Analogously, under the same conditions, MAP_VALUES(<transformation>). FILTER(<predicate>) is equivalent to FILTER(<predicate>).MAP_VA-LUES(<transformation>).

Rule #3: MAP(<expr_list>).FILTER(<predicate1>) is equivalent to FILTER(<predicate2>).MAP(<expr_list>), where predicate1 is rewritten to `predicate2` by substituting `KEY` and `VALUE` as per the mapping defined in `expr_list`. Example:

```
MAP(VALUE[0], KEY).FILTER(KEY > VALUE)  →
FILTER(VALUE[0] > KEY).MAP(VALUE[0], KEY)
```

Since planning a filter as early as possible always increases the efficiency, the planner always takes advantage of moving a filter by applying rules #2 and #3 whenever they are applicable.

4.3 Bind Join

Bind join [10] is an efficient method for implementing semi-joins across heterogeneous data stores that uses subquery rewriting to push the join conditions. In this paper, we adapt the bind join approach for MFR subqueries and we focus on it in our experimental evaluation, as it brings a significant performance gain in certain occasions.

Using bind join between relational data (expressed in an SQL named table expression) and big data (expressed in an MFR named table expression) allows for reducing the computation cost at the DPF and the communication cost between the DPF and the query engine. This approach implicates that the list of distinct values of the join attribute(s) from the relation, preliminarily retrieved from the relational data store, is passed as a filter to the MFR subquery. To illustrate the approach, let us consider the following SELECT statement performing a join between an SQL named table R and an MFR named table H:

```
SELECT H.x, R.y FROM R JOIN H ON R.id = H.id WHERE R.z='abc'
```

To process this query using the bind join method, first, the table R is retrieved from the relational data store; then, assuming that the distinct values of R.id are, $r_1 \ldots r_n$ the condition id IN $(r_1, \ldots, r_n)$ is passed as a FILTER to the MFR subquery that retrieves the dataset H from HDFS data store. Thus, only the tuples from H that match the join criteria are retrieved. Moreover, if the filter condition can be pushed even further in the MFR chain (according to the MFR rewrite rules) and thus to overcome at least one REDUCE operation, this may lead to a significant performance boost, as data will be filtered before at least one shuffle phase.

To estimate the expected performance gain of a bind join, the query optimizer takes into account the overhead a bind join may produce. First, when using bind join, the query engine must wait for the SQL named table to be fully retrieved before initiating the execution of the MFR subquery. Second, if the number of distinct values of the join attribute is large, using a bind join may slower the performance as it requires data to be pushed into the MFR subquery. In the example above, the query engine first asks the RDBMS (e.g. by running an EXPLAIN statement) for an estimation of the cardinality of data retrieved from R, after rewriting the SQL subquery by including the selection condition R.z = 'abc'. If the estimated cardinality does not exceed a certain threshold, the optimizer plans for performing a bind join that can significantly increase the MFR subquery selectivity and affect the volume of transferred data.

5   Use Case Example

In this section, we reveal the steps the query engine takes to process a query using selection pushdown and especially bind join as optimization techniques. We also focus on the way the query engine dynamically rewrites the MFR subquery to perform a bind join. We consider three distinct data stores: PostgreSQL as the relational database (referred to as `rdb`), MongoDB as the document database (referred to as `mongo`) which will be subqueried by SQL expressions that are mapped by the wrapper to MongoDB calls, and an HDFS cluster (referred to as `hdfs`) processed using the Apache Spark framework.

Datasets. For the use case walkthrough we consider small sample datasets in the context of the multistore query example described in Sect. 1.

The `rdb` database stores structured data about scientists and their affiliations in the following table:

Scientists:

Name	Affiliation	Country
Ricardo	UPM	Spain
Martin	CWI	Netherlands
Patrick	INRIA	France
Boyan	INRIA	France
Larri	UPC	Spain
Rui	INESC	Portugal

The `mongo` database contains a document collection about publications including their keywords as follows:

```
Publications(
{ title:'Snapshot Isolation in Cloud DBs',        author:'Ricardo',
            keywords: ['transaction', 'cloud'] },
{ title:'Principles of Distributed Cloud DBs', author:'Patrick',
            keywords: ['cloud', 'storage'] },
{ title:'Graph Databases', author:'Larri', keywords: ['graph', 'NoSQL']}
)
```

HDFS stores unstructured log data from a scientific forum in text files where a single record corresponds to one post and contains a timestamp and username followed by a variable number of fields storing the keywords mentioned in the post:

```
Posts (date, author, kw₁, kw₂, ..., kwₙ)
2014-11-10, alice, storage, cloud
2014-11-10, bob, cloud, virtual, app
2014-11-10, alice, cloud
```

Query 1. This query aims at finding appropriate reviewers for publications of authors with a certain affiliation. It considers each publication's keywords and the experts who have mentioned them most frequently on the scientific forum. The query combines data from the three data stores and can be expressed as follows.

```
scientists( name string, affiliation string )@rdb = (
  SELECT name, affiliation
  FROM scientists
)

publications(autor string, title string, keywords array)@mongo = (
  SELECT author, title, keywords
  FROM publications
)

experts(kw string, expert string)@hdfs = {*
  SCAN(TEXT, 'posts.txt', ',')                                    (op1)
    .FLAT_MAP( lambda data: product(data[2:], [data[1]]) )        (op2)
    .MAP( TUPLE, 1 )                                              (op3)
    .REDUCE( SUM )                                                (op4)
    .MAP( KEY[0], (KEY[1], VALUE) )                               (op5)
    .REDUCE( lambda a, b: b if b[1] > a[1] else a )               (op6)
    .PROJECT(KEY, VALUE[0])                                       (op7)
 *}

SELECT p.author, p.title, e.kw, e.expert
FROM scientists s, publications p, experts e
WHERE s.affiliation = 'INRIA'
  AND p.author = s.name
  AND e.kw IN p.keywords
```

Query 1 contains three subqueries. The first two subqueries is a typical SQL statement to get data about respectively scientists (from PostgreSQL) and scientific publications (from MongoDB). The third subquery is an MFR operation chain that transforms the unstructured log data from the forum posts and represents the result of text analytics as a relation that maps each keyword to the person who has most frequently mentioned it. To achieve the result dataset, the MFR operations request transformations over the stored data, each of which is expressed either in a declarative way or with anonymous (lambda) Python functions.

The SCAN operation **op1** reads data from the specified text source and splits each line to an array of values. Let us recall that the produced array contains the author of the post in its second element and the mentioned keywords in the subarray starting from the third element. The following FLAT_MAP operation **op2** consumes each emitted array as a tuple and transforms each tuple using the defined Python lambda function, which performs a Cartesian product between the keywords subarray and the author, thus emitting a number of keyword-author pairs. Each of these pairs is passed to the MAP operation **op3**, which produces a new dataset, where each keyword-author pair is mapped to a value of 1. Then the REDUCE operation **op4** aggregates the number of occurrences for each keyword-author pair. The next MAP operation **op5** transforms the dataset by mapping each keyword to a pair of author-occurrences. The REDUCE **op6** finds for each keyword the author with the maximum number of occurrences, thus finding the expert who has mostly used the keyword. Finally, the PROJECT defines the mapping between the dataset fields and the columns of the returned relation.

Query Processing. First, Query 1 is compiled into the preliminary execution plan, depicted in Fig. 2. Then, the query optimizer finds the opportunity for pushing down the condition affiliation = 'INRIA' into the relational data store. Thus, the selection condition is included in the WHERE clause of the subquery for rdb. Doing this, the compiler determines that the column s.affiliation is no longer referenced in the common execution plan, so it is simply removed from the corresponding projection on scientists from rdb. This pushdown implies increasing the selectivity of the subquery, which is identified by the optimizer as an opportunity for performing a bind join. To further verify this opportunity, the query optimizer asks rdb to estimate the cardinality for the rewritten SQL subquery and, considering also the availability of an index on the field author in the MongoDB collection publications, the optimizer plans for bind join by pushing into the sub-plan for MongoDB the selection condition author IN <authors>, where <authors> refers to the list of distinct values of the s.name column, which will be determined at runtime.

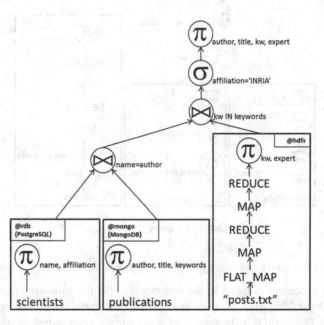

Fig. 2. Preliminary query plan for Query 1

Analogously, by using the catalog information provided by the MongoDB wrapper to estimate the cardinality of the join between scientists and publications, the optimizer plans to also involve the MFR subquery into a bind join and thus pushes the bind join condition kw IN (<keywords>). Here, <keywords> is a placeholder for the list of distinct keywords retrieved from the column p.keywords. Recall that each value in p.keywords is an array, so the query processor will have to first flatten the intermediate relation by transforming the array-type column p.keywords to a scalar-type column named __keywords. Since p.keywords participates in the join condition kw IN keywords, its flattening leads to transforming the join to an equi-join which allows for the query engine to utilize efficient methods for equi-joins.

Furthermore, the MFR planner seeks for opportunities to move the bind join filter condition kw IN (<keywords>) earlier in the MFR operation chain by applying the

MFR rewrite rules, explained below. At this stage, although `<keywords>` is not known, the planner has all the information needed to apply the rules. After these transformations, the optimized query plan (Fig. 3) is executed by the query processor. In this notation, we use the symbol **F** to denote the flattening operator.

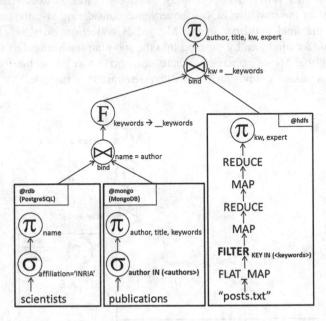

Fig. 3. Optimized query plan for Query 1

To execute the query plan, the query engine takes the following steps:

1. The query processor delivers to the wrapper of `rdb` the following SQL statement, rewritten by taking into account the pushed selection condition, for execution against the PostgreSQL data store, and waits for the corresponding result set to be retrieved in order to compose the bind join condition for the next step.

```
SELECT name
FROM scientists
WHERE affiliation = 'INRIA'
```

2. The MongoDB wrapper prepares a native query to send to the MongoDB database to retrieve those tuples from `publications` that match the bind join criteria. It takes into account the bind join condition derived from the already retrieved data from `rdb` and generates a MongoDB query whose SQL equivalent would be the following:

```
SELECT title, author, keywords FROM publications
WHERE author IN ('Patrick', 'Boyan')
```

However, the wrapper does not generate an SQL statement; instead it generates directly the corresponding MongoDB native query:

```
db.publications.find(
  { author: {$in:['Patrick', 'Boyan']} },
  { title: 1, author: 1, keywords: 1, _id: 0 }
)
```

Upon receiving the result dataset (a MongoDB document collection), the wrapper converts it to a table, according to the signature of the named table expression publications, ready to be joined with the already retrieved result set from step 1. The result of the bind join is the contents of the following intermediate relation:

author	title	keywords
Patrick	Principles of DDBS	['cloud', 'storage']

3. The flattening operator transforms the intermediate relation from step 2 to the following one:

author	title	__keywords
Patrick	Principles of DDBS	cloud
Patrick	Principles of DDBS	storage

4. The query processor identifies a list of the distinct values of the join attribute __keywords and derives from it the bind join condition kw IN ('cloud', 'storage') to push inside the subquery against hdfs.

5. The MFR planner for the wrapper of hdfs decides at which stage of the MFR sequence to insert the filter, by applying a number of rewrite rules. According to rule #1, the planner initially inserts the filter just before the PROJECT **op7** by rewriting the condition expression as follows:

```
.FILTER( KEY IN ('cloud', 'storage') )
```

Next, by applying consecutively rules #2 and #3, the planner moves the FILTER before the MAP **op5** by rewriting its condition expression according to rule #3:

```
.FILTER( KEY[0] IN ('cloud', 'storage') )
```

Analogously, rules #2 and #3 are applied again, moving the FILTER before **op3**, rewriting the expression once again, and thus settling it to its final position. After all transformations the MFR subquery is converted to the final MFR expression below.

```
SCAN( TEXT, 'posts.txt', ',' )
.FLAT_MAP( lambda data: product(data[2:], [data[1]]) )
.FILTER( TUPLE[0] IN ('cloud', 'storage') )
.MAP( TUPLE, 1 )
.REDUCE( SUM )
.MAP( KEY[0], (KEY[1], VALUE) )
.REDUCE( lambda a, b: b if b[1] > a[1] else a )
```

6. The wrapper interprets the reordered MFR sequence, translates it to the Python script below as per the Python API methods of Spark, and executes it within the Spark framework.

```
sc.textFile('posts.txt').map( lambda line: line.split(',') ) \
.flatMap( lambda data: product(data[2:], [data[1]]) ) \
.filter( lambda tup: tup[0] in ['cloud','storage'] ) \
.map( lambda tup: (tup, 1) ) \
.reduceByKey( lambda a, b: a + b ) \
.map( lambda tup: (tup[0][0], (tup[0][1], tup[1])) ) \
.reduceByKey( lambda a, b: b if b[1] > a[1] else a )
```

The result of MFR query reordering and interpreting on Spark is another intermediate relation:

kw	expert
cloud	alice
storage	alice

7. The intermediate relations from steps 3 and 6 are joined to produce the final result that lists the suggested experts for each publication regarding the given keywords:

author	title	kw	expert
Patrick	Principles of DDBS	cloud	alice
Patrick	Principles of DDBS	storage	alice

6 Experimental Validation

The goal of our experimental validation is to evaluate the impact of query rewriting and optimization on execution time. More specifically, we explore the performance benefit of using bind join under different conditions. To achieve this, we have implemented a prototype of our query engine, aiming at implementing the proposed optimization techniques. In this section, we first describe the current implementation of the query engine prototype. Then, we introduce the datasets, based on the use case example in Sect. 5. Finally, we present our experimental results.

6.1 Prototype

For the purpose of our experiments, we have developed a prototype that invokes the Spark SQL [2] engine to perform data integration. The query compiler/optimizer is implemented in C ++; it compiles a CloudMdsQL query into an optimized query execution plan. Then, a flow of invocations of Spark SQL's Python API methods is generated out of the execution plan. Thus, each MFR subquery, after being translated to a Python piece of code, is natively executed in the Spark context, while for performing relational operations on MFR and SQL named tables our prototype takes advantage of Spark SQL's DataFrame API. Wrappers are implemented as Python classes, whose execute() method accepts a native query or a query sub-plan, executes the corresponding query against its data store, and returns a DataFrame object ready to be consumed by relational operators at Spark SQL. In our evaluation scenario, we use three data stores (rdb, mongo, and hdfs) whose wrappers are implemented as follows:

- The PostgreSQL wrapper loads a PostgreSQL data source by invoking `sqlContext.read().format("jdbc")`. Thus, the wrapper is able to execute SQL statements against the relational database using its JDBC driver. The wrapper exports an `explain()` function that the query optimizer invokes to get an estimation of the cost of a subquery. It can also be queried about the existence of certain indexes on table columns and their types.
- The wrapper for MongoDB is implemented as a wrapper to an SQL compatible data store, i.e. it performs native MongoDB query invocations according to their SQL equivalent. It uses the `pymongo` library to query the database and then transforms a result set into a Spark DataFrame. The wrapper maintains the catalog information by running probing queries such as `db.collection.stats()` to keep actual database statistics. Similarly to the PostgreSQL wrapper, it also provides information about available indexes on document attributes.
- The MFR wrapper implements an MFR planner to optimize MFR expressions in accordance with any pushed down selections. The wrapper uses Spark's Python API, and thus translates each transformation to Python lambda functions. Besides, it also accepts raw Python lambda functions as transformation definitions. The wrapper executes the dynamically built Python code using the reflection capabilities of Python by means of the `eval()` function. Then, it transforms the resulting RDD into a Spark DataFrame.

Normally, if the QEP involves no bind joins, after all data frames that correspond to all named tables within a query are loaded into the Spark SQL context, the query engine simply invokes `sqlContext.sql()` to execute the integration SELECT statement as is. In case of a bind join, the query engine takes a couple of more steps. First, it performs a SELECT DISTINCT query on an intermediate table and then uses the retrieved distinct values to build the bind join condition that will be pushed inside the subquery for the other named table that participates in the join. If there is a flatten operator, the query engine uses the LATERAL VIEW clause available in Spark SQL. In our use case example, the publications named table is flattened into a temporary table using the command:

```
SELECT author, title, __keywords
FROM publications
LATERAL VIEW explode(keywords) _k AS __keywords
```

Then, to do the bind join, SELECT DISTINCT __keywords is performed on that temporary table.

6.2 Datasets

We performed our experimental evaluation in the context of the use case example, presented in Sect. 5. For this purpose, we generated data to populate the PostgreSQL table `scientists`, the MongoDB document collection `publications`, and text files with unstructured log data stored in HDFS. All data is uniformly distributed and consistent. The datasets have the following characteristics:

- Table scientists contains 10 K rows, distributed over 1000 distinct affiliations, making 10 authors per affiliation.
- Collection publications contains 10 M documents, with uniform distribution of values of the author attribute, making 1 K publications per scientist. Each publication is randomly assigned a set of 6 to 10 keywords out of 10 K distinct keyword values. Also, there is an association between authors and keywords, so that all the publications of a single author reference only 1 % of all the keywords. This means that a join involving the publications of a single author will have a selectivity factor of 1 %; hence 100 distinct values for the bind join condition. The total size of the collection is 10 GB.
- HDFS contains 16 K files distributed between the nodes, with 100 K tuples per file making 1.6 billion tuples, corresponding to posts from 10 K forum users with 10 K distinct keywords mentioned by them. The first field of each tuple is a timestamp and does not have an impact on the experimental results. The second field contains the author of the post as a string value. The remainder of the tuple line contains 1 to 10 keyword string values, randomly chosen out of the same set of 10 K distinct keywords. The total size of the data is 124 GB.

6.3   Experimental Results

To evaluate the impact of optimization on query execution, we use a cluster of the GRID5000 platform (www.grid5000.fr), with one node for PostgreSQL and MongoDB and 4 to 16 nodes for the HDFS cluster. The Spark cluster, used as both the DPF and the query processor, is collocated with the HDFS cluster. Each node in the cluster runs on 16 CPU cores at 2.4 GHz, 64 GB main memory, and the network bandwidth is 10Gbps.

To demonstrate in detail the optimization techniques and their impact on the query execution, we prepared 3 different queries. We execute each of them in three different HDFS cluster setups – with 4, 8, and 16 nodes. Then we compare the execution times without and with bind join to the MFR subquery, which are illustrated in each query's corresponding graphical chart. We do not focus on evaluating the bind join between PostgreSQL and MongoDB, as its benefit is less significant when compared to the benefit of doing bind join to the MFR subquery, because of the big difference in data sizes.

All the queries use the following common named table expressions, which we created as stored expressions:

```
CREATE NAMED EXPRESSION
scientists( name string, affiliation string )@rdb = (
  SELECT name, affiliation
  FROM scientists
);

CREATE NAMED EXPRESSION
publications(autor string, title string, keywords array)@mongo = (
  SELECT author, title, keywords
  FROM publications
);

CREATE NAMED EXPRESSION
experts(kw string, expert string)@hdfs = {*
  SCAN(TEXT, 'posts.txt', ',')
    .FLAT_MAP( lambda data: product(data[2:], [data[1]]) )
    .MAP( TUPLE, 1 )
    .REDUCE( SUM )
    .MAP( KEY[0], (KEY[1], VALUE) )
    .REDUCE( lambda a, b: b if b[1] > a[1] else a )
    .PROJECT(KEY, VALUE[0])
*};

CREATE NAMED EXPRESSION
experts_alt(kw string, expert string)@hdfs = {*
  SCAN(TEXT, 'posts.txt', ',')
    .FLAT_MAP( lambda data: product(data[2:], [data[1]]) )
    .MAP_VALUES(lambda v: Counter([v]))
    .REDUCE(lambda C1, C2: C1 + C2)
    .MAP_VALUES( lambda C: \
          reduce(lambda a,b: b if b[1] > a[1] else a, C.items()) )
    .PROJECT(KEY, VALUE[0])
*};
```

Thus, each of the queries is expressed as a single SELECT statement that uses the above named table expressions. The named tables scientists, publications, and experts have exactly the same definition as in the use case example from Sect. 5.

The named table experts_alt does the same as experts, but its MFR sequence contains only one REDUCE (respectively, it does only one shuffle) and more complex map functions. It uses Python's Counter dictionary collection, with the additive property to sum up numeric values grouped by the key. The first MAP_VALUES maps a keyword to a Counter object, initialized with a single author key. Then the REDUCE sums all Counter objects associated to a single keyword, so that the result from it is an aggregated Counter dictionary, where an author is mapped to a number of occurrences of the keyword. The final MAP_VALUES uses Python's reduce() function (note that this is not Spark's reduce operator) to choose from all items in a Counter the author with the highest number of occurrences for a keyword.

Query 0 involves only the MongoDB database and the DPF to find experts for the publications of only one author. Thus, the selectivity factor of the bind join is 1 %, as the number of keywords used by a single author is 1 % of the total number of keywords. As we experimented with different number of nodes, we observe that the query execution efficiency and the benefit of the bind join scale well when the number of nodes increases. This is also observed in the rest of the queries.

```
-- Query 0
SELECT p.author, p.title, e.kw, e.expert
FROM publications p, experts e
WHERE p.author = 'author1'
   AND e.kw IN p.keywords
```

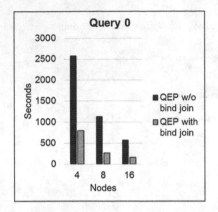

Query 1, as already introduced in Sect. 5, involves all the data stores and aims at finding experts for publications of authors with a certain affiliation. This makes a selectivity factor of 10 % for the bind join, as there are 10 authors per affiliation. In addition, we explore another variant of the query, filtered to three affiliations, or 30 % selectivity factor of the bind join. We enumerate the two variants as Query 1.1 and Query 1.2.

```
-- Query 1.1: selectivity factor 10%
SELECT p.author, p.title, e.kw, e.expert
FROM scientists s, publications p, experts e
WHERE s.affiliation = 'affiliation1'
   AND p.author = s.name AND e.kw IN p.keywords
```

```
-- Query 1.2: selectivity factor 30%
SELECT p.author, p.title, e.kw, e.expert
FROM scientists s, publications p, experts e
WHERE s.affiliation IN ('affiliation1', 'affiliation2', 'affiliation3')
   AND p.author = s.name AND e.kw IN p.keywords
```

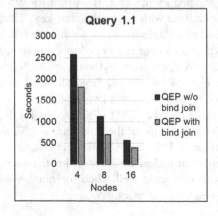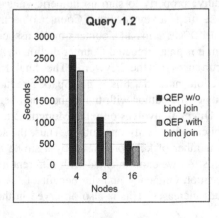

Query 2 does the same as Query 1, but uses the MFR subquery `experts_alt`, which uses more sophisticated map functions, but makes only one shuffle, where the key is a keyword. For comparison, the MFR expression `experts` makes two shuffles, of which the first one uses a bigger key, composed of a keyword-author pair. Therefore, the corresponding Spark computation of Query 2 involves much smaller size of data to be shuffled compared to Query 1, which explains its better overall efficiency and higher relative benefit of using bind join. Like with Query 1, we explore two variants with different selectivity factors of the bind join condition.

```
-- Query 2.1: selectivity factor 10%
SELECT p.author, p.title, e.kw, e.expert
FROM scientists s, publications p, experts_alt e
WHERE s.affiliation = 'affiliation1'
  AND p.author = s.name AND e.kw IN p.keywords
```

```
-- Query 2.2: selectivity factor 30%
SELECT p.author, p.title, e.kw, e.expert
FROM scientists s, publications p, experts_alt e
WHERE s.affiliation IN ('affiliation1', 'affiliation2', 'affiliation3')
  AND p.author = s.name AND e.kw IN p.keywords
```

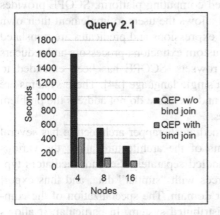 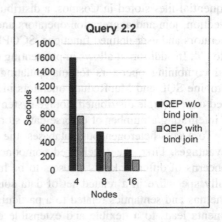

This experimental evaluation illustrates the query engine's ability to perform optimization and choose the most efficient execution plan. The results show the significant benefit of performing bind join in our experimental scenario, despite the overhead it produces (see Sect. 4.3).

7   Related Work

The problem of accessing heterogeneous data sources has long been studied in the context of multidatabase and data integration systems [16]. The typical solution is to provide a common data model and query language to transparently access data sources through a mediator, thus hiding data source heterogeneity and distribution.

The main requirements for a common query language (and data model) are support for nested queries, schema independence, and data-metadata transformation [22]. Nested queries allow queries to be arbitrarily chained together in sequences, so the result of one query (for one data store) may be used as the input of another (for another data store). Schema independence allows the user to formulate queries that are robust in front of schema evolution. Data-metadata transformation is important to deal with heterogeneous data models. To satisfy these requirements, several functional SQL-like languages have been introduced, with Functional SQL [20] being the first of them. More recently, FunSQL [3] has been proposed for the cloud, to allow shipping the code of an application to its data.

With respect to combining SQL and map/reduce operators, a number of SQL-like query languages have been recently introduced. HiveQL is the query language of the data warehousing solution Hive, built on top of Hadoop MapReduce [18]. Hive gives a relational view of HDFS stored unstructured data. HiveQL queries are decomposed to relational operators, which are then compiled to MapReduce jobs to be executed on Hadoop. In addition, HiveQL allows custom scripts, defining MapReduce jobs, to be referred in queries and used in combination with relational operators. SCOPE [6] is a declarative language from Microsoft designed to specify the processing of large sequential files stored in Cosmos, a distributed computing platform. SCOPE provides selection, join and aggregation operators and allows the users to implement their own operators and user-defined functions. SCOPE expressions and predicates are translated into C#. In addition, it allows implementing custom extractors, processors and reducers and combining operators for manipulating rowsets. SCOPE has been extended to combine SQL and MapReduce operators in a single language [24]. These systems are used over a single distributed storage system and therefore do not address the problem of integrating a number of diverse data stores.

To access heterogeneous databases, the mediator/wrapper architecture has several advantages. First, the specialized components of the architecture allow the various concerns of different kinds of users to be handled separately. Second, mediators typically specialize in a related set of data sources with "similar" data, and thus export schemas and semantics related to a particular domain. The specialization of the components leads to a flexible and extensible distributed system. In particular, it allows seamless integration of different data stored in very different data sources, ranging from full-fledged relational databases to simple files. DISCO [19] is a data integration system for accessing Web data sources, using an operator-based approach. It combines a generic cost model with specific cost information provided by the data source wrappers, thus allowing flexible cost estimation.

More recently, with the advent of cloud databases and big data processing frameworks, multidatabase solutions have evolved towards multistore systems that provide integrated access to a number of RDBMS, NoSQL and HDFS data stores through a common query engine. We can divide multistore systems between loosely-coupled, tightly-coupled and hybrid.

Loosely-coupled multistore systems are reminiscent of multidatabase systems in that they can deal with autonomous data stores, which can then be accessed through the multistore system common interface as well as separately through their local API. Most loosely-coupled systems support only read-only queries. Loosely-coupled multistore systems follow the mediator/wrapper architecture with several data stores (e.g. NoSQL and RDBMS). BigIntegrator [14] integrates data from cloud-based NoSQL big data stores, such as Google's Bigtable, and relational databases. The system relies on mapping a limited set of relational operators to native queries expressed in GQL (Google Bigtable query language). With GQL, the task is achievable because it represents a subset of SQL. However, it only works for Bigtable-like systems and cannot integrate data from HDFS. QoX [17] integrates data from RDBMS and HDFS data stores through an XML common data model. It produces SQL statements for relational data stores, and Pig/Hive code for interfacing Hadoop to access HDFS data. The QoX optimizer uses a dataflow approach for optimizing queries over data stores, with a black box approach for cost modeling. SQL ++ [15] mediates SQL and NoSQL data sources through a semi-structured common data model. The data model supports relational operators and to handle efficiently nested data, it also provides a flatten operator. The common query engine translates subqueries to native queries to be executed against data stores with or without schema. All these approaches mediate heterogeneous data stores through a single common data model. The polystore BigDAWG [9] goes one step further by defining "islands of information", where each island corresponds to a specific data model and its language and provides transparent access to a subset of the underlying data stores through the island's data model. The system enables cross-island queries (across different data models) by moving intermediate datasets between islands in an optimized way.

Tightly-coupled multistore systems have been introduced with the goal of integrating Hadoop or Spark for big data analysis with traditional (parallel) RDBMSs. Tightly-coupled multistore systems trade autonomy for performance, typically in a shared-nothing cluster, taking advantage of massive parallelism. Odyssey [11] enables storing and querying data within HDFS and RDBMS, using opportunistic materialized views. MISO [13] is a method for tuning the physical design of a multistore system (Hive/HDFS and RDBMS), i.e. deciding in which data store the data should reside, in order to improve the performance of big data query processing. The intermediate results of query execution are treated as opportunistic materialized views, which can then be placed in the underlying stores to optimize the evaluation of subsequent queries. JEN [23] allows joining data from two data stores, HDFS and RDBMS, with parallel join algorithms, in particular, an efficient zigzag join algorithm, and techniques to minimize data movement. As the data size grows, executing the join on the HDFS side appears to be more efficient. Polybase [8] is a feature of Microsoft SQL Server Parallel Data Warehouse to access HDFS data using SQL. It allows HDFS data to be referenced through external PDW tables and joined with native PDW tables using SQL queries. HadoopDB [1] provides Hadoop MapReduce/HDFS access to multiple single-node RDBMS servers (e.g. PostgreSQL or MySQL) deployed across a cluster, as in a shared-nothing parallel DBMS. It interfaces MapReduce with RDBMS through database connectors that execute SQL queries to return key-value pairs. Estocada [5] is a self-tuning multistore platform for providing access to datasets in native format while automatically placing fragments of the datasets across heterogeneous stores. For query optimization, Estocada combines both cost-based and rule-based approaches.

Hybrid systems support data source autonomy as in loosely-coupled systems, and exploit the local data source interface as in tightly-coupled systems, and typically

HDFS through a parallel data processing framework like MapReduce or Spark. Spark SQL [2] is a parallel SQL engine built on top of Apache Spark and designed to provide tight integration between relational and procedural processing through a declarative API that integrates relational operators with procedural Spark code, taking advantage of massive parallelism. Spark SQL provides a DataFrame API that can map to relations arbitrary object collections and thus enables relational operations across Spark's RDDs and external data sources. In addition, it includes a flexible and extensible optimizer that supports operator pushdowns to data sources, according to their capabilities.

Our work fits in the hybrid system category as, similarly to Spark SQL, it uses Spark API to access the DPF data store, while querying the other stores through an SQL wrapper. However, it adds value by allowing the ad-hoc usage of user-defined map/reduce operators directly in MFR subqueries, yet allowing for optimization through the use of bind join and operator reordering. Furthermore, it does not give up the underlying data store's autonomy.

8 Conclusion

In this paper, we proposed a functional SQL-like query language and query engine to integrate data from relational, NoSQL, and big data stores (such as HDFS). Our query language can directly express subqueries that can take full advantage of the functionality of the underlying data stores and processing frameworks. Furthermore, it allows for query optimization, so that the query operator execution sequence specified by the user may be reordered by taking into account the properties of map/filter/reduce operators together with the properties of relational operators. Finally, compared with the related work on multistore systems, our work fits in the hybrid system category. However, it does not give up data store's autonomy, thus making our approach more general.

Our validation demonstrates that the proposed query language achieves the following requirements. First, it provides high expressivity by allowing the ad-hoc usage of specific map/filter/reduce operators through the MFR notation, as it was demonstrated with the hdfs subqueries. Second, it is optimizable as was demonstrated through performing bind join by rewriting the MFR subquery after retrieving the dataset from the MongoDB database. Finally, it allows for reducing the amount of processed data during the execution of the MFR sequence by reordering MFR operators according to the determined rules. Our performance evaluation illustrates the query engine's ability to optimize a query and choose the most efficient execution strategy.

Acknowledgements. This research has been partially funded by the European Commission under project CoherentPaaS (FP7-611068).

References

1. Abouzeid, A., Badja-Pawlikowski, K., Abadi, D., Silberschatz, A., Rasin, A.: HadoopDB: an architectural hybrid of MapReduce and DBMS technologies for analytical workloads. PVLDB **2**, 922–933 (2009)
2. Armbrust, M., Xin, R., Lian, C., Huai, Y., Liu, D., Bradley, J., Meng, X., Kaftan, T., Franklin, M., Ghodsi, A., Zaharia, M.: Spark SQL: relational data processing in Spark. In: ACM SIGMOD International Conference on Management of Data, pp. 1383–1394 (2015)
3. Binnig, C., Rehrmann, R., Faerber, F., Riewe, R.: FunSQL: it is time to make SQL functional. In: EDBT/ICDT Conference, pp. 41–46 (2012)
4. Bondiombouy, C., Kolev, B., Levchenko, O., Valduriez, P.: Integrating big data and relational data with a functional SQL-like query language. In: Chen, Q., Hameurlain, A., Toumani, F., Wagner, R., Decker, H. (eds.) DEXA 2015. LNCS, vol. 9261, pp. 170–185. Springer, Heidelberg (2015)
5. Bugiotti, F., Bursztyn, D., Deutsch, A., Ileana, I., Manolescu, I.: Invisible glue: scalable self-tuning multi-stores. In: *CIDR* Conference (2015)
6. Chaiken, R., Jenkins, B., Larson, P., Ramsey, B., Shakib, D., Weaver, S., Zhou, J.: SCOPE: easy and efficient parallel processing of massive data sets. PVLDB **1**, 1265–1276 (2008)
7. CoherentPaaS project. http://coherentpaas.eu
8. DeWitt, D., Halverson, A., Nehme, R., Shankar, S., Aguilar-Saborit, J., Avanes, A., Flasza, M., Gramling, M.: Split query processing in Polybase. In: ACM SIGMOD Conference, pp. 1255–1266 (2013)
9. Duggan, J., Elmore, A.J., Stonebraker, M., Balazinska, M., Howe, B., Kepner, J., Madden, S., Maier, D., Mattson, T., Zdonik, S.: The BigDAWG polystore system. ACM SIGMOD Rec. **44**(2), 11–16 (2015)
10. Haas, L., Kossmann, D., Wimmers, E., Yang, J.: Optimizing queries across diverse data sources. In: International Conference on Very Large Databases (VLDB), pp. 276–285 (1997)
11. Hacigümüs, H., Sankaranarayanan, J., Tatemura, J., LeFevre, J., Polyzotis, N.: Odyssey: a multi-store system for evolutionary analytics. PVLDB **6**, 1180–1181 (2013)
12. Kolev, B., Valduriez, P., Bondiombouy, C., Jiménez-Peris, R., Pau, R., Pereira, J.: CloudMdsQL: querying heterogeneous cloud data stores with a common language. In: Distributed and parallel databases, pp. 463–503 (2015). http://link.springer.com/article/10.1007%2Fs10619-015-7185-y
13. LeFevre, J., Sankaranarayanan, J., Hacigümüs, H., Tatemura, J., Polyzotis, N., Carey, M.: MISO: souping up big data query processing with a multistore system. In: ACM SIGMOD Conference, pp. 1591–1602 (2014)
14. Minpeng, Z., Tore, R.: Querying combined cloud-based and relational databases. In: International Conference on Cloud and Service Computing (CSC), pp. 330–335 (2011)
15. Ong, K.W., Papakonstantinou, Y., Vernoux, R.: The SQL ++ semi-structured data model and query language: a capabilities survey of SQL-on-Hadoop, NoSQL and NewSQL databases (2014). Corr, abs/1405.3631
16. Özsu, T., Valduriez, P.: Principles of Distributed Database Systems. Springer, New York (2011)
17. Simitsis, A., Wilkinson, K., Castellanos, M., Dayal, U.: Optimizing analytic data flows for multiple execution engines. In: ACM SIGMOD Conference, pp. 829–840 (2012)
18. Thusoo, A., Sarma, J.S., Jain, N., Shao, Z., Chakka, P., Anthony, S., Liu, H., Wyckoff, P., Murthy, R.: Hive - a warehousing solution over a map-reduce framework. PVLDB **2**, 1626–1629 (2009)

19. Tomasic, A., Raschid, L., Valduriez, P.: Scaling access to heterogeneous data sources with DISCO. IEEE Trans. Knowl. Data Eng. **10**, 808–823 (1998)
20. Valduriez, P., Danforth, S.: Functional SQL, an SQL upward compatible database programming language. Inf. Sci. **62**, 183–203 (1992)
21. Wiederhold, G.: Mediators in the architecture of future information systems. Computer **25**, 38–49 (1992)
22. Wyss, C.M., Robertson, E.L.: Relational languages for metadata integration. ACM Trans. Database Syst. **30**(2), 624–660 (2005)
23. Yuanyuan, T., Zou, T., Özcan, F., Gonscalves, R., Pirahesh, H.: Joins for hybrid warehouses: exploiting massive parallelism in hadoop and enterprise data warehouses. In: EDBT/ICDT Conference, pp. 373–384 (2015)
24. Zhou, J., Bruno, N., Wu, M., Larson, P., Chaiken, R., Shakib, D.: SCOPE: Parallel Databases Meet MapReduce. PVLDB **21**, 611–636 (2012)
25. Zhu, Q., Larson, P.-A.: A query sampling method for estimating local cost parameters in a multidatabase system. In: International Conference on Data Engineering (ICDE), pp. 144–153 (1994)
26. Zhu, Q., Larson, P.-A.: Global query processing and optimization in the CORDS multidatabase system. In: International Conference on Parallel and Distributed Computing Systems, pp. 640–647 (1996)
27. Zhu, Q., Sun, Y., Motheramgari, S.: Developing cost models with qualitative variables for dynamic multidatabase environments. In: International Conference on Data Engineering (ICDE), pp. 413–424 (2000)

Ontology Matching with Knowledge Rules

Shangpu Jiang, Daniel Lowd, Sabin Kafle, and Dejing Dou[⊠]

Department of Computer and Information Science,
University of Oregon, Eugene, USA
{shangpu,lowd,skafle,dou}@cs.uoregon.edu

Abstract. Ontology matching is the process of automatically determining the semantic equivalences between the concepts of two ontologies. Most ontology matching algorithms are based on two types of strategies: terminology-based strategies, which align concepts based on their names or descriptions, and structure-based strategies, which exploit concept hierarchies to find the alignment. In many domains, there is additional information about the relationships of concepts represented in various ways, such as Bayesian networks, decision trees, and association rules. We propose to use the similarities between these relationships to find more accurate alignments. We accomplish this by defining soft constraints that prefer alignments where corresponding concepts have the same local relationships encoded as *knowledge rules*. We use a probabilistic framework to integrate this new *knowledge-based* strategy with standard terminology-based and structure-based strategies. Furthermore, our method is particularly effective in identifying correspondences between complex concepts. Our method achieves better F-score than the state-of-the-art on three ontology matching domains.

1 Introduction

An ontology is an explicit specification of a conceptualization Gruber (1993) in a domain. Ontology matching is the process of aligning two semantically related ontologies in the same domain. Traditionally, this task is performed by human experts from the domain of the ontologies. Since the task is tedious and error prone, especially in large ontologies, there has been substantial work on developing automated or semi-automated ontology matching systems (Shvaiko and Jerome 2013). While some automated matching systems make use of data instances (e.g., Doan et al. (2004)), in this paper we focus on the *schema-level* ontology matching task, in which no data instance is used.

Previous automatic ontology matching systems mainly use two classes of strategies. *Terminology-based* strategies discover corresponding concepts with similar names or descriptions. *Structure-based* strategies discover corresponding groups of concepts with similar hierarchies. In many cases, additional information about the relationships among the concepts is available through domain models, such as Bayesian networks, decision trees, and association rules. A domain model can be represented as a collection of *knowledge rules*, each of which denotes a semantic relationship among several concepts. These relationships may be complex, uncertain, and rely on imprecise numeric values. In this

© Springer-Verlag Berlin Heidelberg 2016
A. Hameurlain et al. (Eds.): TLDKS XXVIII, LNCS 9940, pp. 75–95, 2016.
DOI: 10.1007/978-3-662-53455-7_4

paper, we introduce a new *knowledge-based strategy*, which uses the structure of these knowledge rules as (soft) constraints on the alignment.

As a motivating example, consider two ontologies in the basketball game domain. One ontology has datatype properties `height`, `weight`, `center`, `forward` and `guard` for players, while the other ontology has the corresponding datatype properties `h`, `w`, and `position`. Terminology-based strategies may not identify these correspondences. However, if we know that a large value of `height` implies `center` is true in the first ontology, and the same relationship holds for `h` and `position` = `Center` in the second ontology, then we tend to believe that `height` maps to `h` and `center` maps to `position` = `Center`.

We use Markov logic networks (MLNs) Domingos and Lowd (2009) as a probabilistic language to combine the knowledge-based strategy with other strategies, in a formalism similar to that of Niepert et al. (2010). In particular, we encode the knowledge-based strategy with weighted formulas that increase the probability of alignments where corresponding concepts have isomorphic relationships. We use an MLN inference engine to find the most likely alignment. We name our method Knowledge-Aware Ontology Matching (KAOM).

Our approach is also capable of identifying *complex correspondences*, an extremely difficult task in ontology matching. A complex correspondence is a correspondence between a simple concept and a complex concept (e.g., `grad_student` maps to the union of `PhD` and `Masters`). This can be achieved by constructing a set of *complex concepts* (e.g., unions of concepts) in each ontology, subsequently generating candidate complex correspondences, and using multiple strategies – including the knowledge-based strategy – to find the correct ones.

The contributions of this work are as follows:

- We show how to represent common types of domain models as knowledge rules, and how to use these knowledge rules to obtain more accurate ontology alignments. We combine the knowledge-based strategy with terminological and structural strategies using Markov logic, a coherent probabilistic model.
- By incorporating complex concepts, our approach is also capable of discovering complex correspondences, which is a very difficult scenario in the ontology matching task.
- Our approach is especially effective in identifying the correspondences of numerical or nominal datatype properties as well. This has been very difficult for most schema-level ontology matching approaches.

This paper is an extended version of Jiang et al. (2015). Besides providing more details in background and related work, we have added more ontologies and much more experimental results to show the effectiveness and advantages of our KAOM approach. In the census domain, we added another ontology "pums90" in addition to "adult" and "income", so we have 3 pairs of ontologies for matching instead of 1 pair. We show that our method outperforms the baseline in recall and F1 score in all 3 ontology matching tasks. In the conference domain, we added another 2 ontologies to the original 5, and we have 21 pairs of ontologies instead of 10 pairs. In total, we have added 13 new pairs of ontology matching tasks which basically double the experimental results compared with the conference version of our paper. In general, we show that our method not only outperforms the baseline method that does not use knowledge-based information, but also outperforms two state-of-the-art systems in terms of recall and F1, especially in tasks that automatically discover complex correspondences.

The paper is organized as follows. In Sect. 2, we define ontology matching and review previous work. In Sect. 3, we introduce the concept of "knowledge rules" with a definition and examples. In Sect. 4, we present the knowledge-based strategy. In Sect. 5, we show how to incorporate complex concepts in our method. In Sect. 6, we formalize our method with Markov logic networks. We present experimental results in Sect. 7 and conclude in Sect. 8.

2 Background and Related Work

In AI, an ontology is an explicit specification of a conceptualization (Gruber 1993) in a domain. This conceptualization provides a formal knowledge representation through *concepts* from a domain, and *relationships* between these concepts. The term ontology comes from philosophy, where it corresponds to the study of existence or reality, and as Gruber points out "For knowledge-based systems, what exists is exactly that which can be represented" (Gruber 1993). Through concepts, individuals of these concepts, relationships, and constraints, an ontology provides a vocabulary and a model of the domain it represents. Because of this domain model, it is possible to perform inference. In this work, we consider *Description Logic* based ontologies, as those described through the *Web Ontology Language* (OWL). OWL is the standard ontology language proposed by the *World Wide Web Consortium* (W3C).

In the same domain, people may develop different ontologies or database schemas for their own applications. Therefore, identifying the corresponding entities of different knowledge base or database systems has been a crucial step for semantic integration problems, e.g., ontology translation and ontology merging in the AI community (Noy 2004), and data translation, data integration (Doan and Halevy 2005) and data exchange (Kolaitis 2005) in the database community.

In this work, we will focus on ontologies that are described by OWL. Given two OWL ontologies, a formal definition of ontology matching is given as follows.

Definition 1 (Ontology Matching (Euzenat and Shvaiko 2007)). *Given two ontologies O_1 and O_2, a correspondence is a 3-tuple $\langle e_1, e_2, r \rangle$ where e_1 and e_2 are entities of the first and second ontologies respectively, and r is a semantic relationship such as equivalence $(\equiv)$ and subsumptions $(\sqsubseteq$ or $\sqsupseteq)$. An alignment is a set of correspondences. Ontology matching is the task or process of identifying the correct semantic alignment between the two ontologies. In most cases, ontology matching focuses on equivalence relationships only.*

Automatic or semi-automatic matching discovery has received a lot of attention in recent years in both AI and database communities, such as the approaches described in Shvaiko and Jerome 2013, Rahm and Bernstein 2001. In fact, an ontology and a database schema share many common features, as they both essentially define a vocabulary of concepts/entities and the relationships and constraints among them. Therefore, the approaches for automatic schema matching and ontology matching are similar as well.

Most existing ontology matching and schema matching systems mainly use two types of strategies: terminology-based and structure-based strategies (Shvaiko and Jerome 2013). Terminology-based strategies for ontology matching are based on terminological similarity of concepts and relationships, such as string-based or linguistic similarity measures. Structure-based strategies are based on the assumption that two matching ontologies should have similar local structures, where the structure is represented by subsumption relationships of classes and properties, and domains and ranges of properties. Advanced ontology matching systems often combine the two types of strategies, e.g., PROMPT (Noy and Musen 2000), CUPID (Madhavan et al. 2001), Similarity Flooding (Melnik et al. 2002), iMatch (Albagli et al. 2009) and PRIOR+ (Mao et al. 2010).

Recently, a probabilistic framework based on Markov logic was proposed to combine multiple strategies (Niepert et al. 2010). In particular, it encodes multiple strategies and heuristics into hard and soft constraints of Markov logic, and finds the best matching by minimizing the weighted number of violated constraints. The constraints include string similarity, the cardinality constraints which enforce that each concept matches at most one concept, the coherence constraints which prevent inconsistency induced by the matching, and the stability constraints which penalize dissimilar local subsumption relationships.

Definition 2 (Complex Correspondences). *A complex concept is a composition (e.g., unions, complements) of one or more simple concepts. In OWL[1], there are several constructors for creating complex classes and properties (see the top part of Table 1 for an incomplete list of constructors). A complex correspondence is an equivalence relationship between a simple class or property and a complex class or property in two ontologies* (Ritze et al. 2008).

Previous work has taken several different approaches to find complex correspondences (i.e., complex matching). iMap (Dhamankar et al. 2004) employs a set of searchers, each specialized in certain types of complex correspondences containing operators for primitive classes, such as string concatenation or arithmetic operations on numbers, and uses beam search to reduce the search space. One characteristic of iMap is that it generates explanations for matchings, but it needs domain knowledge to evaluate candidate matchings. Ritze et al. (2008) generate complex correspondences based on linguistic and structural features given a candidate one-to-one alignment. They summarize four patterns of complex correspondences: Class by Attribute Type (CAT), Class by Inverse Attribute Type (CIAT), Class by Attribute Value (CAV), and Property Chain pattern (PC). For example, there are two conditions for a CAT matching pattern $O_1 : a \equiv O_2 : \exists p.b$: a and b are terminologically similar, and the domain of p is a superclass of a.

---

[1] http://www.w3.org/TR/owl2-primer/.

An et al. (2005a; 2005b) provide interesting methods to construct complex mapping rules between relational tables or XML data and ontologies when given an initial set of correspondences between the concepts in the schemas and ontologies. They offer the mapping formalisms to capture the semantics of XML or relational schemas by constructing the semantic trees from them. Their generated rules will be useful to domain experts for further refinement, as well as to applications. Finally, when aligned or overlapping data is available, inductive logic programming (ILP) techniques can be used as well (Hu et al. 2011, Qin et al. 2007).

Many ontology matching or schema matching systems make use of data instances to some extent (Dhamankar et al. 2004, Doan et al. 2002, Hu et al. 2011, Qin et al. 2007). For instance, GLUE (Doan et al. 2002) employs machine learning and exploits data instances to find matchings between concepts. However, in this paper, we focus on the case where data are not available or data sharing is not preferred because of communication cost or privacy concerns.

3 Representation of Domain Knowledge

In the AI community, knowledge is typically represented in *formal languages*, among which ontology-based languages are the most widely used forms. As we mentioned, the Web Ontology Language (OWL) is the W3C standard ontology language that describes the classes and properties of objects in a specific domain. OWL and many other ontology languages are based on variations of description logics. In ontology languages such as OWL, knowledge is represented as logic *axioms*. These axioms describe properties of classes or relationships (e.g., a relationship is functional, symmetric, or antisymmetric, etc.), or a relationship of several entities (e.g., the relationship 'grandfather' is the composition of the two relationships 'father' and 'parent').

The choice of using description logic as the foundation of the Semantic Web ontology languages is largely due to the trade-off between expressivity and reasoning efficiency. In tasks such as ontology matching, reasoning does not need to be instant, so we can afford to consider other forms of knowledge outside of a specific ontology language or description logic.

Definition 3 (Knowledge Rule). *A knowledge rule is a sentence $R(a, b, \ldots; \theta)$ in a formal language which consists of a relationship R, a set of entities (i.e., classes, attributes or relationship) $\{a, b, \ldots\}$, and (optionally) a set of parameters θ. A knowledge rule carries logical or probabilistic semantics representing the relationship among these entities. The specific semantics depend on R.*

Many domain models and other types of knowledge can be represented as sets of knowledge rules, each rule describing the relationship of a small number of entities. The semantics of each relationship R can typically be expressed with a formal language. Table 1 shows some examples of the symbols used in formal languages such as description logic, along with their associated semantics.

Table 1. Syntax and semantics of DL symbols (top), DL axioms (middle), and other knowledge rules used in the examples of the paper (bottom)

Syntax	Semantics
$\top$	$\mathcal{D}$
$\bot$	$\emptyset$
$C \sqcap D$	$C^{\mathcal{I}} \cap D^{\mathcal{I}}$
$C \sqcup D$	$C^{\mathcal{I}} \cup D^{\mathcal{I}}$
$\neg C$	$\mathcal{D} \backslash C^{\mathcal{I}}$
$\forall R.C$	$\{x \in \mathcal{D} \mid \forall y((x,y) \in R^{\mathcal{I}} \to y \in C^{\mathcal{I}})\}$
$\exists R.C$	$\{x \in \mathcal{D} \mid \exists y((x,y) \in R^{\mathcal{I}} \wedge y \in C^{\mathcal{I}})\}$
$R \circ S$	$\{(x,y) \mid \exists z((x,z) \in R^{\mathcal{I}} \wedge (z,y) \in S^{\mathcal{I}})\}$
R^-	$\{(x,y) \mid (y,x) \in R^{\mathcal{I}}\}$
$R \upharpoonright C$	$\{(x,y) \in R^{\mathcal{I}} \mid x \in C^{\mathcal{I}}\}$
$R \downharpoonright C$	$\{(x,y) \in R^{\mathcal{I}} \mid y \in C^{\mathcal{I}}\}$
$C \sqsubseteq D$	$C^{\mathcal{I}} \subseteq D^{\mathcal{I}}$
$C \sqsubseteq \neg D$	$C^{\mathcal{I}} \cap D^{\mathcal{I}} = \emptyset$
$R \prec S$	$y < y'$ for $\forall (x,y) \in R^{\mathcal{I}} \wedge (x,y') \in S^{\mathcal{I}}$
$C \Rightarrow D$	$\Pr(D^{\mathcal{I}} \mid C^{\mathcal{I}})$ is close to 1

We illustrate a few forms of knowledge rules with the following examples. For each rule, we provide a description in English, a logical representation, and an encoding as a knowledge rule with a particular semantic relationship, R_i. We define a new relationship in each example, but, in a large domain model, most relationships would be appear many times in different rules.

Example 1. The submission deadline precedes the camera ready deadline:

$$\texttt{paperDueOn} \prec \texttt{manuscriptDueOn}$$

This is represented as $R_1(\texttt{paperDueOn}, \texttt{manuscriptDueOn})$ with $R_1(a,b) : a \prec b$.

Example 2. A basketball player taller than 81 in. and heavier than 245 pounds is likely to be a center:

$$\texttt{h} > 81 \wedge \texttt{w} > 245 \Rightarrow \texttt{pos} = \texttt{Center}$$

This rule can be viewed as a branch of a *decision tree* or an *association rule*. It can be represented as $R_2(\texttt{h}, \texttt{w}, \texttt{pos=Center}, [81, 245])$, with $R_2(a,b,c,\theta) : a > \theta_1 \wedge b > \theta_2 \Rightarrow c$.

Example 3. A smoker's friend is likely to be a smoker as well:

$$\texttt{Smokes}(x) \wedge \texttt{Friend}(x,y) \Rightarrow \texttt{Smokes}(y)$$

Relational rules such as this one describe relationships of attributes across multiple tables, as opposed to propositional data mining rules that are restricted to a single table. This rule can be represented as $R_3(\texttt{Smokes}, \texttt{Friend})$ with $R_3(a, b) : a(x) \land b(x, y) \Rightarrow a(y)$.

For the remainder of this paper, we will assume that the knowledge in both ontologies is represented as knowledge rules, as described in this section.

4   Our New Knowledge-Based Strategy

We propose a new strategy for ontology matching that uses the similarity of knowledge rules in the two ontologies. It is inspired by the structure-based strategy in many ontology matching algorithms (e.g., (Melnik et al. 2002 and Niepert et al. 2010)). It naturally extends the subsumption relationship of entities in structure-based strategies to other types of relationships.

We use Markov logic to combine the knowledge-based strategy with other strategies. In particular, each strategy is represented as a set of *soft constraints*, each of which assigns a score to the alignments satisfying it, and the alignment with the highest total score is chosen as the best alignment. We now describe the soft constraints encoding the knowledge-based strategy. Our complete Markov logic-based approach, including the soft constraints required for the other strategies, will be described in Sect. 6.

For each relationship R_k that appears in both domains, we introduce a set of soft constraints so that the alignments that preserve these relationships are preferred to those that do not:

$$
\begin{array}{ll}
+w_k & R_k(a, b) \land \neg R_k(a', b') \Rightarrow a \not\equiv a' \lor b \not\equiv b' \\
+w'_k & R_k(a, b) \land R_k(a', b') \Rightarrow a \equiv a' \land b \equiv b' \\
& \hspace{2cm} \forall a, b \in O_1, a', b' \in O_2
\end{array}
$$

These formulas assume R_k is a binary relationship, but they trivially generalize to any arity, e.g., $R_k(a, b, c, d, e, \ldots)$. Note that separate constraints are created for each possible tuple of constants from the respective domains. The numbers preceding the constraints (w_k and w'_k) are the *weights*. A larger weight represents a stronger constraint, since alignments are ranked based on the total weights of the constraints they satisfy. A missing weight means the constraint is a hard constraint which must be satisfied.

Example 4. A reviewer of a paper cannot be the paper's author. In the cmt[2] ontology we have $R_4(\texttt{writePaper}, \texttt{readPaper})$ and in the confOf ontology we have $R_4(\texttt{write}, \texttt{reviews})$ where $R_4(a, b) : a \sqsubseteq \neg b$ is the disjoint relationship of properties. Applying the constraint formulas defined above, we increase the score of all alignments containing the two correct correspondences: writePaper $\equiv$ writes and readPaper $\equiv$ reviews.

---

[2] Throughout the paper, we will use ontologies in the conference domain (cmt, confOf, conference, edas, ekaw) and the NBA domain (nba-os, yahoo) in our examples. The characteristics of these ontologies will be further described in Sect. 7.

Rules involving continuous numerical attributes often include parameters (e.g., thresholds in Example 2) that do not match between different ontologies. In order to apply the knowledge-based strategy to numerical attributes, we make the assumption that corresponding numerical attributes roughly have a *positive linear* transformation. This assumption is often true in real applications, for instance, when an imperial measure of height matches to a metric measure of height. We propose two methods to handle numerical attributes.

The first method is to compute a *distance measure* (e.g., Kullback-Leibler divergence) between the distributions of the corresponding attributes in a candidate alignment. Although the two distributions describe different attributes, the distance can be computed by assuming a linear transformation between the two attributes. The coefficients of the mapping relationship can be roughly estimated using the ranges of attribute values appearing in the knowledge rules (see Example 5 below).

Specifically, if the distance between rules $R(\mathsf{a}, \mathsf{b}, \ldots, \theta)$ and $R(\mathsf{a'}, \mathsf{b'}, \ldots, \theta')$ is d, then we add the constraint:

$$a \equiv a' \wedge b \equiv b' \wedge c \equiv c'$$

with a weight of $\max(d_0 - d, 0)$ for a given threshold d_0.

Example 5. In the nba-os ontology, we have conditional rules converted from a decision tree, such as

$$\mathsf{h} > 81 \wedge \mathsf{w} > 245 \Rightarrow \mathsf{Center}$$

Similarly, in the nbayahoo ontology, we have

$$\mathsf{h'} > 2.06 \wedge \mathsf{w'} > 112.5 \Rightarrow \mathsf{Center'}$$

Here the knowledge rules represent the conditional distributions of multiple entities. We define the distance between the two conditional distributions as

$$d(\mathsf{h}, \mathsf{w}, \mathsf{Center}; \mathsf{h'}, \mathsf{w'}, \mathsf{Center'}) = \mathbb{E}_{p(\mathsf{h}, \mathsf{w})} d(p(\mathsf{Center}|\mathsf{h}, \mathsf{w}) || p(\mathsf{Center'}|\mathsf{h'}, \mathsf{w'}))$$

where $\mathbb{E}(\cdot)$ is expectation and $d(p||p')$ is a distance measure. Because Center and Center' are binary attributes, we simply use $|p - p'|$ as the distance measure. For numerical attributes, we can use the difference of two distribution histograms as the distance measure. We assume the attribute correspondences (h and h', w and w') are linear mappings, and the linear relation can be roughly estimated (e.g., by simply matching the minimum and maximum numbers in these rules). When computing the expectation over h and w, we apply the linear mapping to generate corresponding values of h' and w', e.g., h' = 0.025 h, w' = 0.45 w. The distribution of the conditional attributes $p(\mathsf{h}, \mathsf{w})$ can be roughly estimated as independent and uniform over the ranges of the attributes.

The second method for handling continuous attributes is to discretize them, reducing the continuous attribute problem to the discrete problem described earlier. For example, suppose each continuous attribute x is replaced with a discrete attribute x^d, indicating the quartile of x rather than its original value. Then we have $R_5(\mathtt{h}^d, \mathtt{w}^d, \mathtt{Center})$ and $R_5(\mathtt{h'}^d, \mathtt{w'}^d, \mathtt{Center'})$ with relationship $R_5(a, b, c) : a = 4 \wedge b = 4 \Rightarrow c$, and the discrete value of 4 indicates that both a and b are in the top quartile. Other discretization methods are also possible, as long as the discretization is done the same way in both domains.

Our method does not rely on the forms of knowledge rules, nor does it rely on the algorithms used to learn these rules. As long as similar techniques or tools are used on both sides of ontologies, we would always be able to find interesting knowledge-based similarities between the two ontologies.

5   Finding Complex Correspondences

Our approach can also find complex correspondences, which contain complex concepts in either or both of the ontologies. We add the complex concepts into consideration and treat them the same way as simple concepts. Then we jointly solve all the simple and complex correspondences by considering terminology, structure, and knowledge-based strategies in a single probabilistic formulation.

First, because complex concepts may be recursively defined and potentially infinite, we need to select a finite subset of complex concepts and use them to generate the candidate correspondences. We will only include the complex concepts occurring in the ontology axioms or in the knowledge rules.

Second, we need to define a string similarity measure for each type of complex correspondence. For example, Ritze et al. (2008) requires two conditions for a Class by Attribute Type (CAT) matching pattern $O_1 : a \equiv O_2 : \exists p.b$ (e.g., $a =$ Accepted_Paper, $p = $ hasDecision, $b = $Acceptance): a and b are terminologically similar, and the domain of p (Paper in the example) is a superclass of a. We can therefore define the string similarity of a and $\exists p.b$ to be the string similarity of a and b which coincides with the first condition, and the second condition is encoded in the structure stability constraints. The string similarity measure of many other types of correspondences can be defined similarly based on the heuristic method in Ritze et al. (2008). If there does not exist a straight-forward way to define the string similarity for a certain type of complex correspondences, we can simply set it to 0 and rely on other strategies to identify such correspondences.

Lastly, we need constraints for the correspondence of two complex concepts. The corresponding component concepts and same constructor always implies the corresponding complex concepts, while in the other direction, it is a soft constraint.

$$\mathrm{cons}_k(a, b) \equiv \mathrm{cons}_k(a', b') \Leftarrow a \equiv a' \wedge b \equiv b'$$
$$+w_k^c \qquad \mathrm{cons}_k(a, b) \equiv \mathrm{cons}_k(a', b') \Rightarrow a \equiv a' \wedge b \equiv b'$$

where cons_k are different constructors for complex concepts, e.g., union, $\exists p.b$.

Some complex correspondences are almost impossible to be identified with traditional strategies. With the knowledge-based strategy, it becomes possible.

Example 6. A reviewer of a paper cannot be the paper's author. In the `cmt` ontology we have

$$\texttt{writePaper} \sqsubseteq \neg\texttt{readPaper}$$

and in the `conference` ontology we have

$$\texttt{contributes} \mathbin{\lfloor} \texttt{Reviewed\_contribution} \sqsubseteq \neg(\texttt{contributes} \circ \texttt{reviews})$$

We first build two complex concepts `contributes` ⌊ `Reviewed_contribution` and `contributes` ∘ `reviews`. With $R_4(a,b) = a \sqsubseteq \neg b$ (disjoint properties), the score function would favor the correspondences

$$\texttt{writePaper} \equiv \texttt{contributes} \mathbin{\lfloor} \texttt{Reviewed\_contribution}$$
$$\texttt{readPaper} \equiv \texttt{contributes} \circ \texttt{reviews}$$

6 Knowledge Aware Ontology Matching

In this section, we present our approach, Knowledge Aware Ontology Matching (KAOM). KAOM uses Markov logic networks (MLNs) to solve the ontology matching task. The MLN formulation is similar to Nieper et al. (2010) but incorporates the knowledge-based matching strategy and treatment of complex correspondences.

An MLN (Domingos and Lowd 2009) is a set of weighted formulas in first-order logic. Given a set of constants for individuals in a domain, an MLN induces a probability distribution over Herbrand interpretations or "possible worlds". In the ontology matching problem, we represent a correspondence in first-order logic using a binary relationship, `match(a1,a2)`, which is true if concept `a1` from the first ontology is semantically equivalent to concept `a2` from the second ontology (e.g., `match(writePaper, writes)` means `writePaper` ≡ `writes`). Each possible world therefore corresponds to an alignment of the two ontologies. We want to find the most probable possible world, which is the configuration that maximizes the sum of weights of satisfied formulas.

We define three components of the MLN of the ontology matching problem: *constants*, *evidence* and *formulas*. The logical constants are the entities in both ontologies, including the simple named ones and the complex ones. The evidence includes the complete set of OWL-supported relationships (e.g., subsumptions and disjointness) among all concepts in each ontology, and rules represented as first-order atomic predicates as described in the Sect. 3. We use an OWL reasoner to create the complete set of OWL axioms.

For the formulas, we begin with a set of formulas adapted from Nieper et al. (2010):

1. *A-priori similarity* is the string similarity between all pairs of concepts:

$$s_{a,a'} \quad \texttt{match}(a,a')$$

where $s_{a,a'}$ is the string similarity between a and a', which also serves as the weight of the formula. We use the Levenshtein measure (Levenshtein 1966) for simple correspondences. This atomic formula increases the probability of matching pairs of concepts with similar strings, all other things being equal.

2. *Cardinality constraints* enforce one-to-one simple (or complex) correspondences:

$$\texttt{match}(a, a') \wedge \texttt{match}(a, a'') \Rightarrow a' = a''$$

3. *Coherence constraints* enforce consistency of subclass relationships:

$$\texttt{match}(a, a') \wedge \texttt{match}(b, b') \wedge a \sqsubseteq b \Rightarrow a' \sqsubseteq \neg b'$$

4. *Stability constraints* enforce consistency of the subclass relationships between the two ontologies. They can be viewed as a special case of the knowledge-based constraints we introduce below.

Knowledge-Based Constraints. We now describe how we incorporate knowledge-based constraints into the MLN formulation through new formulas relating knowledge rules to matchings. The *stability* constraints in Nieper et al. (2010) consider three subclass relationships, including a is a subclass of b (`subclass`), and a is a subclass or superclass of the domain or range of a property b (`domainsub`, `rangesub`). We extend the relationships (knowledge rule patterns) to sub-property, disjoint properties, and user-defined relationships such as ordering of dates, and non-deterministic relationships such as correlation and anti-correlation:

$$-w_k \quad R_k(a, b, \ldots) \wedge \neg R_k(a', b', \ldots) \Rightarrow \texttt{match}(a, a') \wedge \texttt{match}(b, b') \wedge \ldots, k = 1, \ldots, m \tag{1}$$

where m is the number of knowledge rule patterns. User-defined relationships include those derived from decision trees, association rules, expert systems, and other knowledge sources outside the ontology.

Besides the stability constraints, we introduce a new group of *similarity* constraints that encourage knowledge rules with the same pattern to have corresponding concepts.

$$+w'_k \quad R_k(a, b, \ldots) \wedge R_k(a', b', \ldots) \Rightarrow \texttt{match}(a, a') \wedge \texttt{match}(b, b') \wedge \ldots, k = 1, \ldots, m \tag{2}$$

For numerical rules, we instead use MLN formulas:

$$d_0 - d \quad \texttt{match}(a, a') \wedge \texttt{match}(b, b') \wedge \ldots, k = 1, \ldots, m \tag{3}$$

where d is a distance measure of the two rules $R_k(a, b, \ldots)$ and $R'_k(a', b', \ldots)$ and d_0 is a threshold determining whether the rules are similar or not.

To handle complex correspondences, we add complex concepts that occur in knowledge rules as constants of the MLN, and add knowledge rules that contain these new complex concepts. We define the string similarity and enforce type constraints between simple and complex concepts, as described in Sect. 5. For complex to complex correspondences, the string similarity measure is zero, but we have constraints

$$
w_k^c \qquad
\begin{aligned}
&\mathtt{match}(a, a') \wedge \mathtt{match}(b, b') \wedge \ldots \Rightarrow \mathtt{match}(c, c') \\
&\mathtt{match}(a, a') \wedge \mathtt{match}(b, b') \wedge \ldots \Leftarrow \mathtt{match}(c, c')
\end{aligned}
$$

where $c = \mathrm{cons}_k(a, b, \ldots), c' = \mathrm{cons}_k(a', b', \ldots)$ for each constructor cons_k.

7   Experiments

We test our KAOM approach on three domains: NBA, census, and conference. The sizes of the ontologies of these domains are listed in Table 2. These domains contain very different forms of ontologies and knowledge rules, so we can examine the generality and robustness of our approach.

We use Pellet (Sirin et al. 2007) for logical inference of the ontological axioms and TheBeast[3] (Riedel 2008) and Rockit[4] (Noessner et al. 2013) for Markov logic inference. We ran all experiments on a machine with 24 Intel Xeon E5-2640 cores @2500 MHz and 8GB memory. We compare our system (KAOM) with three others: KAOM without the knowledge-based strategy (MLOM), CODI (Huber et al. 2011) (the same set of formulas as MLOM with a different MLN

Table 2. Number of classes, object properties, data properties and nominal values of each ontology used in the experiments.

domain	ontology	# classes	# object props	# data props	# values
NBA	nba-os	3	3	20	3
	yahoo	4	4	21	7
census	adult	1	0	15	101
	income	1	0	12	97
	pums90	1	0	11	93
OntoFarm	cmt	36	50	10	0
	confOf	38	13	25	0
	conference	60	46	18	6
	edas	103	30	20	0
	ekaw	78	33	0	0
	iasted	133	38	3	0
	sigkdd	45	17	11	0

---

[3] http://code.google.com/p/thebeast/.

[4] https://code.google.com/p/rockit/. We use RockIt for the census domain because TheBeast is not able to handle the large number of rules in that domain.

implementation), and logmap2 (Jiménez-Ruiz et al. 2012), a top performing system in OAEI 2014[5].

We manually specify the weights of the Markov logic formulas in KAOM and MLOM. The weights of stability constraints for subclass relationships are set with values same as the ones used in (Niepert et al. 2010), i.e., the weight for subclass is -0.5, and those for sub-domain and range are -0.25. In KAOM, we also set the weights for different types of similarity rules based on our assessment of their relative importance and kept these weights fixed during the experiments.

7.1 NBA

The NBA domain is a simple experiment we use to demonstrate the effectiveness of our approach. We collected data from the NBA official website and the Yahoo NBA website. For each ontology, we used the WinMine toolkit[6] to learn a decision tree for each attribute using the other attributes as inputs.

For each pair of conditional distributions based on decision tree with up to three attributes, we calculate their similarity based on the distance measure described in Example 5. We use the Markov logic formula (3) with the threshold $d_0 = 0.2$. To make the task more challenging, we did not use any name similarity measures. Our method successfully identified the correspondence of all the numerical and nominal attributes, including height, weight and positions (center, forward and guard) of players. In contrast, without a name similarity measure, no other method can solve the matching problem at all.

7.2 Census

We consider three census datasets and their ontologies from UC Irvine data repository[7]. All three datasets represent census data but are sampled and post-processed differently. These census ontologies are flat with a single concept but many datatype properties and nominal values. For this domain, we use association rules as the knowledge. We first discretize each numerical attribute into five intervals, and then generate association rules for each ontology using the Apriori algorithm with a minimum confidence of 0.9 and minimum support of 0.001. For example, one generated rule is:

```
age='(-inf-25.5]' education='11th' hours-per-week='(-inf-35.5]'
  ==> adjusted-gross-income='<=50K' conf:(1)
```

This is represented as

$$R_6(\text{age}^d, 11\text{th}, \text{hours-per-week}^d, \text{adjusted-gross-income}^d)$$

where x^d refers to the discretized value of x, split into one fifth percentile intervals, and $R_6(a, b, c, d) : a = 1 \land b \land c = 1 \Rightarrow d = 1$. For scalability reasons, we

---

[5] http://oaei.ontologymatching.org/2014/.

[6] http://research.microsoft.com/en-us/um/people/dmax/WinMine/Tooldoc.htm.

[7] https://archive.ics.uci.edu/ml/datasets.html.

consider up to three concepts in a knowledge rule, i.e., association rules with up to three attributes. The weight of knowledge-based constraints are chosen to balance with the string similarities. For this experiment we set it to 0.01.

In Nieper et al. (2010), only the correspondences with apriori similarity measure larger than a threshold τ are added as evidence. We set τ with different values from 0.50 to 0.90. When τ is large, we deliberately discard the string similarity information for some correspondences. Our baseline MLOM for this task is an extension of Nieper et al. (2010) by adding correspondences of *nominal values* and their dependencies with the related attributes. The results are shown in Figs. 1, 2 and 3.

We can see that MLOM outperforms KAOM in terms of precision, while KAOM always gets better recall and F1-score in all three ontology matching tasks. This means our approach fully leverages the knowledge rule information

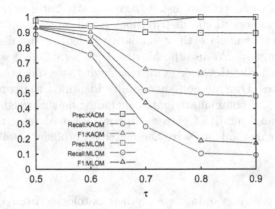

Fig. 1. Precision, recall and F1 on the census domain (`adult` and `income` ontologies) as a function of the string similarity threshold τ.

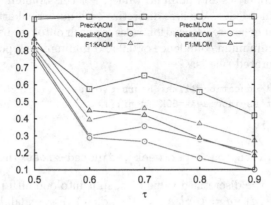

Fig. 2. Precision, recall and F1 on the census domain (`adult` and `pums90` ontologies) as a function of the string similarity threshold τ

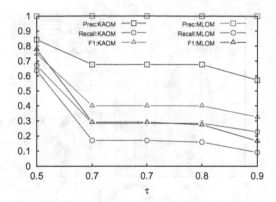

Fig. 3. Precision, recall and F1 on the census domain (`pums90` and `income` ontologies) as a function of the string similarity threshold τ

and thus does not rely too much on the names of the concepts to determine the matching. For example, in the `adult` and `income` pair, when τ is 0.70, KAOM finds 6 out of 8 correspondences of values of `adult:workclass` and `income:class_of_worker`, while MLOM finds none.

For ontology matching task between `pums90` and `adult` ontologies and between `pums90` and `income`, the corresponding names are even more different. Yet, KAOM is always successful in finding the mapping between the attribute `yearsch` present in `pums90` ontology and the attribute `education` in both `adult` and `income` ontologies. Unsurprisingly, MLOM is also able to obtain mappings between attributes `pob` (place of birth) and `native-country` present in `pums90` and `adult` ontology due to the large number of matches in the nominal values of both attributes. The difference between the attribute mappings of group (`education`, `yearsch`) and (`pob`, `native-country`) is in the nominal values of the attributes, with the former having dissimilar nominal value sets while the latter has the exact nominal value sets. This clearly indicate the advantage of KAOM over MLOM in such cases.

The other two systems, CODI and logmap2, were not designed for nominal value correspondences. For instance, in the `adult` and `income` ontology matching, CODI only finds 7 and logmap2 only finds 3 attribute correspondences, while KAOM and MLOM find all the 12 attribute correspondences.

7.3 OntoFarm

In order to show how our system can use manually created expert knowledge bases, we use OntoFarm, a standard ontology matching benchmark for an academic conference domain as the third domain in our experiments. As part of OAEI, it has been widely used in the evaluation of ontology matching systems. We have used 7 of the OntoFarm ontologies (`cmt`, `conference`, `confOf`, `edas`, `ekaw`, `iasted`, `sigkdd`) for this experiment. Using their knowledge of computer

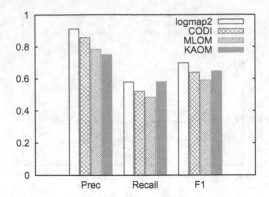

Fig. 4. Average precision, recall and F1 on the OntoFarm domain with only the one-to-one correspondences

science conferences and the structure of just one ontology, two individuals (i.e., human experts) listed a number of rules (e.g., Example 1). We then translated these rules into each of the 7 ontologies. Thus, the same knowledge was added to each of the ontologies, but its representation depended on the specific ontology. For some ontologies, some of the rules were not representable with the concepts in them and thus had to be omitted. This manually constructed knowledge base was developed before running any experiments and kept fixed throughout our experiments. Among the 7 ontologies, we have 21 pairs of matching tasks in total. We set τ to 0.70, and the weight for the knowledge similarity constraints to 1.0. It is hard to show 21 matching results in figures. We show the average precision, recall, and F1 measures of 21 matching results from different methods.

We first compare the four methods to the reference one-to-one alignment from the benchmark (Fig. 4). KAOM has worse precision than the state-of-the-art systems such as logmap2 and CODI, but has comparable or better recall. It was able to identify correspondences in which the concept names are very different, for instance, `cmt:readPaper` ≡ `confOf:reviews`. Note that the similarity constraints work in concert with other constraints. For instance, in Example 4, since disjointness is a symmetric knowledge rule, domain and range constraints could be helpful to identify whether `cmt:writePaper` should match to `confOf:writes` or `confOf:reviews`.

To evaluate our approach on complex correspondences, we extended the reference alignment with hand-labeled complex correspondences (Fig. 5). MLOM does not perform well in this task because the complex correspondences require a good similarity measure to become candidates (such as the linguistic features in Ritze et al. (2008)). KAOM, however, uses the structure of the rules to find many complex correspondences without relying on complex similarity measures. KAOM also outperforms the logmap2 and CODI in recall and F1, despite that we do not use any complex linguistic features but merely the Levenshtein similarity.

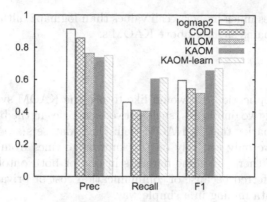

Fig. 5. Average precision, recall and F1 on the OntoFarm domain with the complex correspondences.

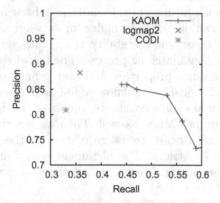

Fig. 6. Average precision-recall curve on the OntoFarm domain with the complex correspondences.

For this task we also tried learning the weights of the formulas[8] (KAOM-learn). For each pair (i.e., 2 ontologies) of the 21 pairs of ontologies, we used the other 5 ontologies (i.e., 7 in total) as training data. So there are 10 pairs in the 5 ontologies for training. KAOM-learn performs slightly better than KAOM.

With the hand-picked or automatically learned weights, KAOM produces a single most-likely alignment. However, we can further tune KAOM to produce alignments with higher recall or higher precision. We accomplish this by adding the MLN formula $\mathtt{match}(a, a')$ with weight w. When w is positive, alignments with more matches are more likely, and when w is negative, alignments with fewer matches are more likely (all other things being equal). We adjusted this weight to produce the precision-recall curve shown in Fig. 6. KAOM dominates

---

[8] We used MIRA implemented in TheBeast for weight learning.

CODI and provides much higher recall values than logmap2, although logmap2's best precision remains slightly above KAOM's.

7.4 Discussion

In a real world application, we would like to use the KAOM system in the following way. Given two ontologies, such as two ontologies in the business domain, there are two scenarios that fit KAOM well: (1) If datasets are not available in either ontology, we only can rely on the knowledge rules from the ontologies, themselves. (2) If there are some datasets in one or both ontologies, but data sharing is not preferred because of communication cost or privacy concerns, we still can utilize data mining in a simple way.

In the first scenario, to map two ontologies, we can use Pellet to automatically create the complete set of OWL axioms, and then manually represent those OWL axioms in MLN as the evidence. The translation from OWL axioms to MLN can be automatic, in such future work, because it is basically a syntax translation process. Then we can use MLN formulas for A-priori similarity, cardinality constraints, coherence constraints, stability constraints, and knowledge-based constraints for the ontology matching process. The knowledge-based constraints include sub-property, disjoint properties, and user-defined relationships. Representing user-defined relationships as knowledge-based constraints is a manual process. If any ontology has some constants, such as named entities or relationships, we will add them into MLN, as well. Then we can run KAOM, with the support of TheBeast and Rockit, to assign a score to the potential alignment that satisfies each soft constraint. The alignment with the highest total score is chosen as the best alignment. We output all alignments with scores that are higher than a threshold (e.g., 0.8).

In the second scenario, if one or both ontologies may have datasets, we can use WinMine to generate decision tree rules and association rules. Both rules can be manually represented as knowledge-based constraints in MLN. Again, this process can be automatic with a (future) syntax translator; but the selection of generated decision tree rules or association rules requires a human to make some judgments. Therefore, it cannot be fully automatic. The other steps, such as constants, evidence, and constraints from ontologies themselves, are generated in the same way as in the first scenario. The knowledge-based constraints generated from data mining provide additional knowledge for KAOM. Then we can run KAOM, with the support of TheBeast and Rockit, to assign a score to any potential alignment that satisfies each soft constraint. The alignment with the highest total score is chosen as the best alignment. We output all alignments with scores that are higher than the threshold.

Of note: We need to manually specify the weights of the Markov logic formulas in KAOM, in the ways explained in the beginning of this section. Although the experiments are from different domains, we can see that the performance of KAOM in the second scenario (i.e., NBA and Census) is better than in the first scenario (i.e., OntoFarm), which does not have any dataset. This is a tradeoff between performance and data availability. Even without any dataset, the performance of KAOM in OntoFarm is still satisfiable. In real-world applications, such as in the business domain, it is very possible that one or both ontologies would have some datasets. KAOM can take advantage of the data without data sharing, which most machine learning-based matching approaches, e.g., GLUE, require. The system can be scaled to larger ontologies (i.e., those with more concepts) than the reported ones, because adding more soft constraints will not make the inference much harder. If the real world application is extended to database schema matching, we believe the foreign keys or other integrity constraints can be presented as knowledge-based constraints in MLN, as well, considering that the database constraints can be represented as logic rules. Therefore, KAOM can be applied to database schema matching without any change.

8 Conclusion

We proposed a new ontology matching algorithm KAOM. The key component of KAOM is the knowledge-based strategy, which is based on the intuition that ontologies about the same domain should contain similar knowledge rules, in spite of the different terminologies they use. KAOM is also capable of discovering complex correspondences, by treating complex concepts the same way as simple ones. We encode the knowledge-based strategy and other strategies in Markov logic and find the best alignment with its inference tools. Experiments on the datasets and ontologies from three different domains show that our method effectively uses knowledge rules of different forms to outperform several state-of-the-art ontology matching methods.

Acknowledgement. This research is being funded by NSF grant IIS-1118050. The authors would like to thank the generous comments from the anonymous reviewers. Their comments have greatly helped improve this research and prepare this paper.

References

OWL Web Ontology Language. http://www.w3.org/TR/owl-ref/

Albagli, S., Shimony, S., Ben-Eliyahu-Zohary, R.: Markov network based ontology matching. In: Proceedings of the 21st International Joint Conference on Artificial Intelligence (IJCAI 2009) (2009). https://www.aaai.org/ocs/index.php/IJCAI/IJCAI-09/paper/view/442/831

An, Y., Borgida, A., Mylopoulos, J.: Inferring complex semantic mappings between relational tables and ontologies from simple correspondences. In: Meersman, R., Tari, Z. (eds.) OTM 2005. LNCS, vol. 3761, pp. 1152–1169. Springer, Heidelberg (2005a). doi:10.1007/11575801_15

An, Y., Borgida, A., Mylopoulos, J.: constructing complex semantic mappings between XML data and ontologies. In: Gil, Y., Motta, E., Benjamins, V.R., Musen, M.A. (eds.) ISWC 2005. LNCS, vol. 3729, pp. 6–20. Springer, Heidelberg (2005b). doi:10.1007/11574620_4

Dhamankar, R., Lee, Y., Doan, A., Halevy, A., Domingos, P.: iMAP: discovering complex semantic matches between database schemas. In: Proceedings of the 2004 ACM SIGMOD International Conference on Management of Data, pp. 383–394 (2004). doi:10.1145/1007568.1007612. ISBN:1-58113-859-8

Doan, A., Halevy, A.Y.: Semantic-integration research in the database community. AI Mag. **26**(1), 83–94 (2005). http://dl.acm.org/citation.cfm?id=1090488.1090497, ISSN:0738-4602

Doan, A., Madhavan, J., Domingos, P., Halevy, A.: Learning to map between ontologies on the semantic web. In: Proceedings of the 11th International Conference on World Wide Web, pp. 662–673 (2002). doi:10.1145/511446.51153, ISBN:1-58113-449-5

Doan, A., Madhavan, J., Domingos, P., Halevy, A.: Ontology matching: a machine learning approach. In: Staab, S., Studer, R. (eds.) Handbook on Ontologies in Information Systems, pp. 385–403. Springer, New York (2004)

Domingos, P., Lowd, D., Logic, M.: An Interface Layer for Artificial Intelligence. Synthesis Lectures on Artificial Intelligence and Machine Learning. Morgan & Claypool, San Rafael (2009). http://books.google.com/books?id=ijqFfoIy_T0C, ISBN:9781598296921

Euzenat, J., Shvaiko, P.: Ontology Matching. Springer-Verlag New York Inc., Secaucus (2007). ISBN:3540496114

Gruber, T.R.: A translation approach to portable ontology specifications. Knowl. Acquis. **5**(2), 199–220 (1993). doi:10.1006/knac.1993.1008

Hu, W., Chen, J., Zhang, H., Qu, Y.: Learning complex mappings between ontologies. In: Proceedings of Joint International Semantic Technology Conference, pp. 350–357 (2011)

Huber, J., Sztyler, T., Noessner, J., Meilicke, C.: CODI: combinatorial optimization for data integration-results for OAEI 2011. In: Ontology Matching, p. 134 (2011)

Jiang, S., Lowd, D., Dou, D.: Ontology matching with knowledge rules. In: Chen, Q., Hameurlain, A., Toumani, F., Wagner, R., Decker, H. (eds.) DEXA 2015. LNCS, vol. 9262, pp. 94–108. Springer, Heidelberg (2015). doi:10.1007/978-3-319-22849-5_7

Jiménez-Ruiz, E., Grau, B.C., Zhou, Y.: LogMap. 2.0: towards logic-based, scalable and interactive ontology matching. In: Proceedings of the 4th International Workshop on Semantic Web Applications and Tools for the Life Sciences, SWAT4LS 2011, pp. 45–46 (2012). doi:10.1145/2166896.2166911, ISBN:978-1-4503-1076-5

Kolaitis, P.G.: Schema mappings, data exchange, metadata management. In: Proceedings of the Twenty-Fourth ACM SIGMOD-SIGACT-SIGART Symposium on Principles of Database Systems, PODS 2005, pp. 61–75. ACM, New York (2005). doi:10.1145/1065167.1065176, ISBN:1-59593-062-0

Levenshtein, V.: Binary codes capable of correcting deletions, insertions and reversals. Soviet Physics Doklady **10**, 707 (1966)

Madhavan, J., Bernstein, P.A., Rahm, E.: Generic schema matching with Cupid. In: The VLDB Journal, pp. 49–58 (2001)

Mao, M., Peng, Y., Spring, M.: An adaptive ontology mapping approach with neural network based constraint satisfaction. Web Semant. **8**(1), 14–25 (2010). doi:10.1016/j.websem.2009.11.002. ISSN:1570-8268

Melnik, S., Garcia-Molina, H., Rahm, E.: Similarity flooding: a versatile graph matching algorithm. In: Proceedings of Eighteenth International Conference on Data Engineering (2002)

Niepert, M., Meilicke, C., Stuckenschmidt, H.: A probabilistic-logical framework for ontology matching. In: Fox M., Poole, D. (eds.) Proceedings of the 24th AAAI Conference on Artificial Intelligence, pp. 1413–1418, July 2010

Noessner, J., Niepert, M., Stuckenschmidt, H.: RockIt: exploiting parallelism and symmetry for MAP inference in statistical relational models. In Proceedings of the Twenty-Seventh AAAI Conference on Artificial Intelligence (2013). http://www.aaai.org/ocs/index.php/AAAI/AAAI13/paper/view/6240

Noy, N.F.: Semantic integration: a survey of ontology-based approaches. SIGMOD Rec. **33**(4), 65–70 (2004). doi:10.1145/1041410.1041421. ISSN:0163-5808

Noy, N.F., Musen, M.A.: PROMPT: algorithm and tool for automated ontology merging and alignment. In: Proceedings of the Seventeenth National Conference on Artificial Intelligence and Twelfth Conference on Innovative Applications of Artificial Intelligence, pp. 450–455 (2000). http://dl.acm.org/citation.cfm?id=647288.721118, ISBN:0-262-51112-6

Qin, H., Dou, D., LePendu, P.: Discovering executable semantic mappings between ontologies. In: Meersman, R., Tari, Z. (eds.) OTM 2007. LNCS, vol. 4803, pp. 832–849. Springer, Heidelberg (2007). doi:10.1007/978-3-540-76848-7_56

Rahm, E., Bernstein, P.A.: A survey of approaches to automatic schema matching. VLDB J. **10**(4), 334–350 (2001). doi:10.1007/s007780100057. ISSN:1066-8888

Riedel, S.: Improving the accuracy and efficiency of MAP inference for Markov logic. In: Proceedings of the 24th Conference on Uncertainty in Artificial Intelligence (UAI 2008), pp. 468–475 (2008)

Ritze, D., Meilicke, C., Svb-Zamazal, O., Stuckenschmidt, H.: A pattern-based ontology matching approach for detecting complex correspondences. In: Ontology Matching (OM 2009), vol. 551 (2008). http://dblp.uni-trier.de/db/conf/semweb/om2009.html#RitzeMSS08

Shvaiko, P., Jerome, E.: Ontology matching: state of the art and future challenges. IEEE Trans. Knowl. Data Eng. **25**(1), 158–176 (2013). doi:10.1109/TKDE.2011.253. ISSN:1041-4347

Sirin, E., Parsia, B., Grau, B.C., Kalyanpur, A., Katz, Y.: Pellet: a practical OWL-DL reasoner. Web Semant. **5**(2), 51–53 (2007). doi:10.1016/j.websem.2007.03.004. ISSN:1570-8268

Regularized Cost-Model Oblivious Database Tuning with Reinforcement Learning

Debabrota Basu[1], Qian Lin[1], Weidong Chen[1], Hoang Tam Vo[3], Zihong Yuan[1],
Pierre Senellart[1,2(✉)], and Stéphane Bressan[1]

[1] School of Computing, National University of Singapore, Singapore, Singapore
debabrota.basu@u.nus.edu
[2] LTCI, CNRS, Télécom ParisTech, Université Paris-Saclay, Paris, France
pierre.senellart@telecom-paristech.fr
[3] SAP Research and Innovation, Singapore, Singapore

Abstract. In this paper, we propose a learning approach to adaptive performance tuning of database applications. The objective is to validate the opportunity to devise a tuning strategy that does not need prior knowledge of a cost model. Instead, the cost model is learned through reinforcement learning. We instantiate our approach to the use case of index tuning. We model the execution of queries and updates as a Markov decision process whose states are database configurations, actions are configuration changes, and rewards are functions of the cost of configuration change and query and update evaluation. During the reinforcement learning process, we face two important challenges: the unavailability of a cost model and the size of the state space. To address the former, we iteratively learn the cost model, in a principled manner, using regularization to avoid overfitting. To address the latter, we devise strategies to prune the state space, both in the general case and for the use case of index tuning. We empirically and comparatively evaluate our approach on a standard OLTP dataset. We show that our approach is competitive with state-of-the-art adaptive index tuning, which is dependent on a cost model.

1 Introduction

In a recent SIGMOD blog entry [23], Guy Lohman asked "Is query optimization a 'solved' problem?". He argued that current query optimizers and their cost models can be critically wrong. Instead of relying on wrong cost models, Stillger et al. have proposed LEO-DB2, a *learning* optimizer [40]; its enhanced performance with respect to classical query optimizers strengthens the claim of discrepancies introduced by the predetermined cost models. Stillger et al. have proposed LEO-DB2, a *learning* optimizer [40]; its enhanced performance with respect to classical query optimizers strengthens the claim of discrepancies introduced by the predetermined cost models.

This is our perspective in this article: we propose a learning approach to performance tuning of database applications. By performance tuning, we mean selection of an optimal physical database configuration in view of the workload.

A. Hameurlain et al. (Eds.): TLDKS XXVIII, LNCS 9940, pp. 96–132, 2016.
DOI: 10.1007/978-3-662-53455-7_5

In general, configurations differ in the indexes, materialized views, partitions, replicas, and other parameters. While most existing tuning systems and literature [10,38,39] rely on a predefined cost model, the objective of this work is to validate the opportunity for a tuning strategy to do without.

To achieve this, we propose a formulation of database tuning as a *reinforcement learning* problem (see Sect. 3). The execution of queries and updates is modelled as a Markov decision process whose states are database configurations, whose actions are configuration changes, and whose rewards are functions of the cost of configuration change and query/update evaluation. This formulation does not rely on a preexisting cost model, rather it learns it.

We present a solution to the reinforcement learning formulation that tackles the curse of dimensionality (Sect. 4). To do this, we reduce the search space by exploiting the quasi-metric properties of the configuration change cost, and we approximate the cumulative cost with a linear model. We formally prove that, assuming such a linear approximation is sound, our approach converges to an optimal policy for estimating the cost.

We then tackle in Sect. 5 the problem of overfitting: to avoid instability while learning the cost model, we add a regularization term in learning the cost model. We formally derive a bound on the total regret of the regularized estimation, that is logarithmic in the time step (i.e., in the size of the workload).

We instantiate our approach to the use case of index tuning (Sect. 6), developing in particular optimizations specific to this use case to reduce the search space. The approaches of Sects. 4 and 5 provide us with two algorithms COREIL and rCOREIL to solve the index tuning problem.

We use this case to demonstrate the validity of a cost-model oblivious database tuning with reinforcement learning, through experimental evaluation on a TPC-C workload [30] (see Sect. 7). We compare the performance with the *Work Function Index Tuning* (WFIT) algorithm [39]. Results show that our approach is competitive yet does not need knowledge of a cost model. While the comparison of WFIT with COREIL establishes reinforcement learning as an effective approach to automatize the index tuning problem, performance of rCOREIL with respect to COREIL demonstrates that the learning performance is significantly enhanced by a crisp estimation of the cost model.

Related work is discussed in Sect. 2.

This article extends a conference publication [6]. In addition to minor edits and precisions added throughout the paper, the following material is novel: the discussion of related work on reinforcement learning (Sect. 2.2); the result of convergence to an optimal policy (Proposition 2) and its proof; the introduction of regularization (Sect. 5), including bounds on regrets (Theorem 1 and Sect. 6.4), the experimental comparison between regularized and non-regularized versions of our approach (Sect. 7.4) and the study of the quality of their cost estimators (Sect. 7.5).

2 Related Work

We now review the related work in two areas of relevance: self-tuning databases, and the use of reinforcement learning for data management applications.

2.1 Automated Database Configuration

Table 1 provides a brief classification of related research in automated database configuration, in terms of various dimensions: the offline or online nature of the algorithm (see further), and the physical design aspects being considered by these works (index selection, vertical partitioning, or mixed design together with horizontal partitioning and replication).

Offline Algorithms. Traditionally, automated database configuration is conducted in an offline manner. In that approach, database administrators (DBAs) identify representative workloads from the trace of historical database queries and updates. That task can be done either manually or with the help of sophisticated tools provided by database vendors. Based on these representative workloads, new database configurations are realized: for example, new beneficial indexes to be created [1,20,45], smart vertical partitioning for reducing I/O costs [17,22,33], or possibly a combination of index selection, partitioning and replication for both stand-alone databases [11,13,27] and parallel databases [2,32].

Online Algorithms. Given the increasing complication and agility of database applications, coupled with the introduction of modern database environments such as database-as-a-service, the aforementioned manual task of DBAs, though it can be done at regular times in an offline fashion, becomes even more tedious and problematic. Therefore, it is desirable to design more automated solutions to the database design problem that are able to continuously monitor changes in the workload and react in a timely manner by adapting the database configuration to the new workload. In fact, the problem of online index selection has been well-studied in the past [10,12,24,25,39]. Generally, these techniques adopt the same working model in which the system continuously tracks the incoming queries for identifying candidate indexes, profiles the benefit of the indexes, and realizes the ones that are most useful for query execution. Specifically, an online approach to physical design tuning (index selection) was proposed in [10]. The essence of the

Table 1. Automated physical database design

	Index selection	Vert. partitioning	Mixed
Offline	[1,20,45]	[17,22,33]	Stand-alone:[11,13,27]
			Parallel DBs:[2,32]
Online	[10,24,25,39]	[3,21,36]	

algorithm is to progressively choose the optimal plan at each step by using a case-by-case analysis on the potential benefits that we may lose by not implementing relevant candidate indexes. That is, each new database configuration is selected only when a physical change, i.e., creating or deleting an index, would be helpful in improving system performance. Similarly, a framework for continuous online physical tuning was proposed in [38] where effective indexes are created and deleted in response to the shifting workload. Furthermore, the framework is able to self-regulate its performance by providing explicit mechanism for controlling the overhead of profiling the benefit of indexes.

One of the key components of an index selection algorithm is profiling indexes' benefit, i.e., how to evaluate the cost of executing a query workload with the new indexes as well as the cost of configuration transition, i.e., creating and deleting indexes. To realize this function, most of the aforementioned online algorithms exploit a *what-if optimizer* [14] which returns such estimated costs. For examples, the what-if optimizer of DB2 was used in [39], and the what-if optimizer of SQL Server was employed in [10], while the classical optimizer of PostgreSQL was extended to support what-if analysis in [38]. However, it is well-known that invoking the optimizer for estimating the cost of each query under different configurations is expensive [27]. In this work, we propose an algorithm that does not require the use of a what-if optimizer while being able to adaptively provide a better database configuration in the end, as reported in our experimental results.

More recently, as column-oriented databases have attracted a great deal of attention in both academia and industry, online algorithms for automated vertical partitioning becomes critical for such an emerging breed of database systems [3,21,36]. Specifically, a storage advisor for SAP's HANA in-memory database system was proposed in [36] in order to take advantage of both columnar and row-oriented storage layouts. At the core of that storage advisor is a cost model which is used to estimate and compare query execution times for different stores. Similarly, a continuous layout adaptation has been recently been introduced in [3] with the aim to support multiple storage layouts in a single engine which is able to adapt to changing data access patterns. The adaptive store monitors the access patterns of incoming queries through a dynamic window of N queries and devises cost models for evaluating the workload and layout transformation cost. Furthermore, in order to efficiently find a near optimal data layout for a given workload, the hybrid store exploits proper heuristic techniques to prune the immense search space of alternative data layouts. On the contrary, the algorithm introduced in [21] uses data mining techniques for vertical partitioning in database systems. The technique is based on closed item sets mining from a query set and system statistic information at run-time, and hence is able to automatically detect changing workloads and perform a re-partitioning action without the need of interaction from DBAs.

Recently, [5,9] have aimed at solving the problem of online index selection for large join queries and data warehouses. Since both approaches use a pre-defined cost model, similarly to [39], it renders them susceptible to erroneous estimations of the cost model. In our work, we are removing the effects of errors made by the cost model by learning it. This gives our approach more robustness and flexibility than cost-model–dependent ones. Moreover, [9] uses genetic algorithms to optimize the selection of multi-table indexes incrementally. But genetic algorithms generally performs worse than reinforcement learning [34] in this kind of dynamic optimization tasks due to its more exploratory nature. In addition, reinforcement learning agents have a more real-time behaviour than genetic behaviour. In [5], authors use a heuristics approach where they incrementally look for frequent itemsets in a query workload. With the knowledge base acquired from there updates, indexes are generated for the frequent itemsets while eliminating the ones generated for infrequent itemsets. Due to such greedy behaviour, higher variance and instability is expected than for reinforcement learning approaches where a trade-off between exploration and exploitation is reached through learning.

Discussion. Compared to the state-of-the-art in online automated database design, our proposed overall approach is more general and has the potential to be applied for various problems such as index selection, horizontal/vertical partitioning design, and in combination with replication as well; we note however that we experiment solely with index tuning in this article. More importantly, as our proposed online algorithm is able to learn the estimated cost of queries gradually through subsequent iterations, it does not need the what-if optimizer for estimating query cost. Therefore, our proposed algorithm is applicable to a wider range of database management systems which may not implement a what-if optimizer or expose its interface to users.

2.2   Reinforcement Learning in Data Management

Reinforcement learning [41] is about determining *the best thing to do next* under an evolving knowledge of the world, in order to reach a goal. This goal is commonly represented as maximization of the cumulative reward obtained while performing some actions. Here each action leads to an individual reward and to a new state, usually in a stochastic manner. *Markov decision processes* (MDPs) [29] are a common model for reinforcement learning scenarios. In this model each action leads to a new state and to a given reward according to a probability distribution that must be learned. This implies an inherent trade-off between exploration (trying out new actions leading to new states and to potentially high rewards) and exploitation (performing actions already known to yield high rewards), a compromise explored in depth in, e.g., the stateless model of multi-armed bandits [4]. Despite being well-adapted to the modelling of uncertain environments, the use of MDPs in data management applications has been limited so far. The use of MDP for modelling data cleaning tasks has been raised in [7]. In that paper, the authors discussed the absence of a straightforward technique to do that because of the huge state space. More generally, the following reasons may explain the difficulties in using reinforcement learning in data management applications:

(i) As in [7], the state space is typically huge, representing all possible partial knowledge of the world. This can be phrased as the *curse of dimensionality*.
(ii) States have complex structures, namely that of the data, or, in our case, of the database configuration.
(iii) Rewards may be delayed, obtained after a long sequence of state transitions. This is for example the case in focused Web crawling, which is a data-centric application domain. Still multi-armed bandits have been successfully applied [16] to this problem.
(iv) Because of data uncertainty, there may be only *partial observability* of the current state. That changes the problem to a partially observable Markov decision process [43].

The last two issues are fortunately not prevalent in the online tuning problem discussed here. This lets us formulate online tuning problem as an MDP and focus on a solution to the first two problems.

3   Problem Definition

Let R be a logical database schema. We can consider R to be the set of its possible database instances. Let S be the set of corresponding physical database configurations of instances of R. For a given database instance, two configurations s and s' may differ in the indexes, materialized views, partitions, replicas, and other parameters. For a given instance, two configurations will be logically equivalent if they yield the same results for all queries and updates.

The cost of changing configuration from $s \in S$ to $s' \in S$ is denoted by the function $\delta(s, s')$. The function $\delta(s, s')$ is not necessarily symmetric, i.e., we may have $\delta(s, s') \neq \delta(s', s)$. This property emerges as the cost of changing configuration from s to s' and the reverse may not be the same. On the other hand, it is a non-negative function and also verifies the identity of indiscernibles: formally, $\delta(s, s') \geq 0$ and the equality holds if and only if $s = s'$. Physically this means that there is no free configuration change. As it is always cheaper to do a direct configuration change, we get

$$\forall s, s', s'' \in S \quad \delta(s, s'') \leq \delta(s, s') + \delta(s', s'').$$

This is simply the triangle inequality. As δ exhibits the aforementioned properties, it is a quasi-metric on S.

Let Q be a workload, defined as a schedule of queries and updates. For brevity, we refer to both of them as *queries*. To simplify, we consider the schedule to be sequential and the issue of concurrency control orthogonal to the current presentation. Thus, query q_t represents the tth query in the schedule, which is executed at time t.

We model a query q_t as a random variable, whose generating distribution may not be known *a priori*. It means that q_t is only observable at time t.

The cost of executing query $q \in Q$ on configuration $s \in S$ is denoted by the function $cost(s, q)$. For a given query, the *cost* function is always positive as the system have to pay some cost to execute a query.

Let s_0 be the initial configuration of the database. At any time t the configuration is changed from s_{t-1} to s_t with the following events in order:

1. Arrival of query q_t. We call $\hat{q}_t$ the observation of q_t at time t.
2. Choice of the configuration $s_t \in S$ based on $\hat{q}_1, \hat{q}_2, \ldots, \hat{q}_t$ and s_{t-1}.
3. Change of configuration from s_{t-1} to s_t. If no configuration change occurs at time t, then $s_t = s_{t-1}$.
4. Execution of query $\hat{q}_t$ under the configuration s_t.

Thus, the system has to pay the sum of the cost of configuration change and that of query execution during each transition. Now, we define *per-stage cost* as

$$C(s_{t-1}, s_t, \hat{q}_t) := \delta(s_{t-1}, s_t) + cost(s_t, \hat{q}_t).$$

We can phrase in other words the stochastic decision process of choosing the configuration changes as a *Markov decision process* (MDP) [29] where states are database configurations, actions are configuration changes, and penalties (negative rewards) are the per-stage cost of the action. Note that in contrast to the general framework of MDPs, transitions from one state to another on an action are deterministic. Indeed, in this process there is no uncertainty associated with the new configuration when a configuration change is decided. On the other hand, penalties are *stochastic*, as they depend on the query which is a random variable. In the absence of a reliable cost model, the cost of a query in a configuration is not known in advance. This makes penalties *uncertain*.

Ideally, the problem would be to find the sequence of configurations that minimizes the sum of future per-stage costs. We assume an infinite horizon [41], which means an action will affect all the future states and actions of the system. But it makes the cumulative sum of future per-stage costs infinite. One practical way to circumvent this problem is to introduce a *discount factor* $\gamma \in [0, 1)$. Mathematically, it makes the cumulative sum of per-stage costs convergent. Physically, it gives more importance to immediate costs than to costs distant in the future, which is a practical intuition. Now, the problem translates into finding the sequence of configurations that minimize a *discounted cumulative cost* defined with γ. Under Markov assumption, a sequence of configuration changes is represented by a function, called *policy* $\pi \colon S \times Q \to S$. Given the current configuration s_{t-1} and a query $\hat{q}_t$, a policy π determines the next configuration $s_t := \pi(s_{t-1}, \hat{q}_t)$.

We define the *cost-to-go* function V^π for a policy π as:

$$V^\pi(s) := \mathbb{E}\left[\sum_{t=1}^{\infty} \gamma^{t-1} C(s_{t-1}, s_t, \hat{q}_t)\right] \quad \text{such that} \quad \begin{cases} s_0 = s \\ s_t = \pi(s_{t-1}, \hat{q}_t), t \geq 1 \end{cases} \tag{1}$$

where $0 < \gamma < 1$ is the discount factor. The value of $V^\pi(s)$ represents the expected cumulative cost for the following policy π from the current configuration s.

Let $\mathcal{U}$ be the set of all policies for a given database schema. Our problem can now be formally phrased as to minimize the expected cumulative cost, i.e., to find an optimal policy π^* such that

$$\pi^* := \arg\min_{\pi \in \mathcal{U}} V^\pi(s_0)$$

where the initial state s_0 is given.

4   Adaptive Database Tuning

4.1   Algorithm Framework

In order to find the optimal policy π^*, we start from an arbitrary policy π, compute an estimation of its cost-to-go function, and incrementally attempt to improve it using the current estimate of the cost-to-go function $\overline{V}$ for each $s \in S$. This strategy is known as *policy iteration* [41] in the reinforcement learning literature.

Traditionally, policy iteration functions as follows. Assuming the probability distribution of q_t is known in advance, we improve the cost-to-go function $\overline{V}^{\pi_t}$ of the policy π_t at iteration t using

$$\overline{V}^{\pi_t}(s) = \min_{s' \in S} \left(\delta(s, s') + \mathbb{E}\left[cost(s', q)\right] + \gamma \overline{V}^{\pi_{t-1}}(s') \right) \qquad (2)$$

We obtain the updated policy as $\arg\min_{\pi_t \in \mathcal{U}} \overline{V}^{\pi_t}(s)$. The algorithm terminates when there is no change in the policy. The proof of optimality and convergence of policy iteration can be found in [28].

Unfortunately, policy iteration suffers from several problems. First, there may not be any proper model available beforehand for the cost function $cost(s, q)$. Second, the curse of dimensionality [28] makes the direct computation of $\overline{V}$ hard. Third, the probability distribution of queries is not assumed to be known *a priori*, making it impossible to compute the expected cost of query execution $\mathbb{E}\left[cost(s', q)\right]$.

Instead, we apply the basic framework shown in Algorithm 1. The initial policy π_0 and cost model C_0 can be initialized arbitrarily or using some intelligent heuristics. In line 5 of Algorithm 1, we have tried to overcome the issues at the root of the curse of dimensionality by juxtaposing the original problem with

---

Algorithm 1. Algorithm Framework

1: Initialization: an arbitrary policy π_0 and a cost model C_0
2: Repeat till convergence
3:      $\overline{V}^{\pi_{t-1}} \leftarrow$ approximate using a linear projection over $\phi(s)$
4:      $C^{t-1} \leftarrow$ approximate using a linear projection over $\eta(s, \hat{q}_t)$
5:      $\pi_t \leftarrow \arg\min_{s \in S'} \left(C^{t-1} + \gamma \overline{V}^{\pi_{t-1}}(s) \right)$
6: End

---

approximated per-stage cost and cost-to-go function. Firstly, we map a configuration to a vector of associated feature $\phi(s)$. Then, we approximate the cost-to-go function by a linear model $\theta^T \phi(s)$ with parameter θ. It is extracted from a reduced subspace S' of configuration space S that makes the search for optimal policy computationally cheaper. Finally, we learn the per-stage cost $C(s, s', \hat{q})$ by a linear model $\zeta^T \eta(s, \hat{q})$ with parameter ζ. This method does not need any prior knowledge of the cost model, rather it learns the model iteratively. Thus, we have resolved shortcomings of policy iteration and the need of predefined cost model for the performance tuning problem in our algorithm. These methods are depicted and analysed in the following sections.

4.2   Reducing the Search Space

In order to reduce the size of search space in line 5 of Algorithm 1, we filter the configurations that satisfy certain necessary conditions deduced from an optimal policy.

Proposition 1. *Let s be any configuration and $\hat{q}$ be any observed query. Let π^* be an optimal policy. If $\pi^*(s, \hat{q}) = s'$, then $cost(s, \hat{q}) - cost(s', \hat{q}) \geq 0$. Furthermore, if $\delta(s, s') > 0$, i.e., if the configurations certainly change after a query, then $cost(s, \hat{q}) - cost(s', \hat{q}) > 0$.*

Proof. Since $\pi^*(s, \hat{q}) = s'$, we have

$$
\begin{aligned}
\delta(s, s') &+ cost(s', \hat{q}) + \gamma V(s') \\
&\leq cost(s, \hat{q}) + \gamma V(s) \\
&= cost(s, \hat{q}) + \gamma \mathbb{E} \left[\min_{s''} \left(\delta(s, s'') + cost(s'', \hat{q}) + \gamma V(s'') \right) \right] \\
&\leq cost(s, \hat{q}) + \gamma \delta(s, s') + \gamma V(s'),
\end{aligned}
$$

where the second inequality is obtained by exploiting triangle inequality $\delta(s, s'') \leq \delta(s, s') + \delta(s', s'')$, as δ is a quasi-metric on S.

This infers that

$$
cost(s, \hat{q}) - cost(s', \hat{q}) \geq (1 - \gamma)\delta(s, s') \geq 0.
$$

The assertion follows.   □

By Proposition 1, if π^* is an optimal policy and $s' = \pi^*(s, \hat{q}) \neq s$, then $cost(s, \hat{q}) > cost(s', \hat{q})$. Thus, we can define a reduced subspace as

$$
S_{s,\hat{q}} = \{s' \in S \mid cost(s, \hat{q}) > cost(s', \hat{q})\}.
$$

Hence, at each time t, we can solve

$$
\pi_t = \arg\min_{s \in S_{s_{t-1}, \hat{q}_t}} \left(\delta(s_{t-1}, s) + cost(s, \hat{q}_t) + \gamma \overline{V}^{\pi^{t-1}}(s) \right). \tag{3}
$$

Next, we design an algorithm that converges to an optimal policy through searching in the reduced set $S_{s,\hat{q}}$.

4.3   Modified Policy Iteration with Cost Model Learning

We calculate the optimal policy using the *least square policy iteration* (LSPI) [18]. If for any policy π, there exists a vector $\boldsymbol{\theta}$ such that we can approximate $V^\pi(s) = \boldsymbol{\theta}^T \phi(\mathbf{s})$ for any configuration s, then LSPI converges to the optimal policy. This mathematical guarantee makes LSPI a useful tool to solve the MDP as defined in Sect. 3. But the LSPI algorithm needs a predefined cost model to update the policy and evaluate the cost-to-go function. It is not obvious that any form of cost model would be available and as mentioned in Sect. 1, pre-defined cost models may be critically wrong. This motivates us to develop another form of the algorithm, where the cost model can be equivalently obtained through learning.

Assume that there exists a feature mapping η such that $cost(s, q) \approx \boldsymbol{\zeta}^T \boldsymbol{\eta}(s, q)$ for some vector $\boldsymbol{\zeta}$. Changing the configuration from s to s' can be considered as executing a special query $q(s, s')$. Therefore we approximate

$$\delta(s, s') = cost(s, q(s, s')) \approx \boldsymbol{\zeta}^T \boldsymbol{\eta}(s, q(s, s')).$$

The vector $\boldsymbol{\zeta}$ can be updated iteratively using the well-known *recursive least squares estimation* (RLSE) [44] as shown in Algorithm 2, where $\boldsymbol{\eta}^t = \eta(s_{t-1}, \hat{q}_t)$ and $\hat{\epsilon}^t = (\boldsymbol{\zeta}^{t-1})^T \boldsymbol{\eta}^t - cost(s_{t-1}, \hat{q}_t)$ is the prediction error. Combining RLSE with LSPI, we get our cost-model oblivious algorithm as shown in Algorithm 3.

In Algorithm 3, the vector $\boldsymbol{\theta}$ determines the current policy. We can make a decision by solving the equation in line 6. The values of $\delta(s_{t-1}, s)$ and $cost(s, \hat{q}_t)$ are obtained from the approximation of the cost model. The vector $\boldsymbol{\theta}^t$ is used to approximate the cost-to-go function following the current policy. If $\boldsymbol{\theta}^t$ converges, then we update the current policy (line 14–16).

Instead of using any heuristics we initialize policy π_0 as initial configuration s_0 and the cost-model C_0 as 0, as shown in the lines 1–3 of Algorithm 3.

Proposition 2. *If for any policy π, there exists a vector $\boldsymbol{\theta}$ such that $V^\pi(s) = \boldsymbol{\theta}^T \phi(s)$ for any configuration s, Algorithm 3 will converge to an optimal policy.*

Proof. Let $\mathcal{V} \colon S \to \mathbb{R}$ be a set of bounded, real-valued functions. Then $\mathcal{V}$ is a Banach space with the norm $\|v\| = \|v\|_\infty = \sup |v(s)|$ for any $v \in \mathcal{V}$.

Algorithm 2. Recursive least squares estimation.

1: **procedure** RLSE($\hat{\epsilon}^t, \overline{B}^{t-1}, \boldsymbol{\zeta}^{t-1}, \boldsymbol{\eta}^t$)

2:      $\gamma^t \leftarrow 1 + (\boldsymbol{\eta}^t)^T \overline{B}^{t-1} \boldsymbol{\eta}^t$

3:      $\overline{B}^t \leftarrow \overline{B}^{t-1} - \frac{1}{\gamma^t} (\overline{B}^{t-1} \boldsymbol{\eta}^t (\boldsymbol{\eta}^t)^T \overline{B}^{t-1})$

4:      $\boldsymbol{\zeta}^t \leftarrow \boldsymbol{\zeta}^{t-1} - \frac{1}{\gamma^t} \overline{B}^{t-1} \boldsymbol{\eta}^t \hat{\epsilon}^t$

5:      **return** $\overline{B}^t, \boldsymbol{\zeta}^t$.

6: **end procedure**

Algorithm 3. Least squares policy iteration with RLSE.

1: Initialize the configuration s_0.
2: Initialize $\theta^0 = \theta = 0$ and $B^0 = \epsilon I$.
3: Initialize $\zeta^0 = 0$ and $\overline{B}^0 = \varepsilon I$.
4: **for** t=1,2,3,... **do**
5:     Let $\hat{q}_t$ be the just received query.
6:         $s_t \leftarrow \underset{s \in S_{s_{t-1}, \hat{q}_t}}{\arg\min} \; (\zeta^{t-1})^T \eta(s_{t-1}, q(s_{t-1}, s)) + (\zeta^{t-1})^T \eta(s, \hat{q}_t) + \gamma \theta^T \phi(s)$
7:     Change the configuration to s_t.
8:     Execute query $\hat{q}_t$.
9:     $\hat{C}^t \leftarrow \delta(s_{t-1}, s_t) + cost(s_t, \hat{q}_t)$.
10:     $\hat{\epsilon}^t \leftarrow (\zeta^{t-1})^T \eta(s_{t-1}, \hat{q}_t) - cost(s_{t-1}, \hat{q}_t)$
11:     $B^t \leftarrow B^{t-1} - \frac{B^{t-1}\phi(s_{t-1})(\phi(s_{t-1}) - \gamma\phi(s_t))^T B^{t-1}}{1 + (\phi(s_{t-1}) - \gamma\phi(s_t))^T B^{t-1}\phi(s_{t-1})}$.
12:     $\theta^t \leftarrow \theta^{t-1} + \frac{(\hat{C}^t - (\phi(s_{t-1}) - \gamma\phi(s_t))^T \theta^{t-1})B^{t-1}\phi(s_{t-1})}{1 + (\phi(s_{t-1}) - \gamma\phi(s_t))^T B^{t-1}\phi(s_{t-1})}$.
13:     $(\overline{B}^t, \zeta^t) \leftarrow RLSE(\hat{\epsilon}^t, \overline{B}^{t-1}, \zeta^{t-1}, \eta^t)$
14:     **if** (θ^t) converges **then**
15:         $\theta \leftarrow \theta^t$.
16:     **end if**
17: **end for**

If we redefine our problem in the reduced search space, we get:

$$\underset{\pi \in \mathcal{U}}{\arg\min} \; \mathbb{E}\left[\sum_{t=1}^{\infty} \gamma^{t-1} \left(\delta(s_{t-1}, s_t) + cost(s_t, q) \right) \right] \qquad (4)$$

$$\text{such that: } s_t = \pi(s_{t-1}, q), \quad s_t \in S_{s_{t-1}, q}, \quad \text{for } t \geq 1$$

Then Algorithm 3 is analogous to LSPI over the reduced search space. For this new problem given by Eq. (4), Algorithm 3 converges to a unique cost-to-go function $\tilde{V} \in \mathcal{V}$. We need to show that $V^* = \tilde{V}$. That means we need to prove the cost-to-go function estimate by Algorithm 3 is the optimal one.

Let us define the process of updating policy as a mapping $\mathcal{M}: \mathcal{V} \rightarrow \mathcal{V}$. Now based on Eq. (2), it can be expressed as

$$\mathcal{M}v(s) = \mathbb{E}\left[\min_{s' \in S_{s,q}} \left(\delta(s, s') + cost(s', q) + \gamma v(s') \right) \right].$$

For a particular configuration s and query q, let

$$a^*_{s,q}(v) = \arg \min_{s' \in S_{s,q}} \left(\delta(s, s') + cost(s', q) + \gamma v(s') \right).$$

Assume that $\mathcal{M}v(s) \geq \mathcal{M}u(s)$. Then

$$
\begin{aligned}
0 &\leq \mathcal{M}v(s) - \mathcal{M}u(s) \\
&= \mathbb{E}\left[\delta(s, a^*_{s,q}(v)) + cost(a^*_{s,q}(v), q) + \gamma v(a^*_{s,q}(v))\right] \\
&\quad - \mathbb{E}\left[\delta(s, a^*_{s,q}(u)) + cost(a^*_{s,q}(u), q) + \gamma u(a^*_{s,q}(u))\right] \\
&\leq \mathbb{E}\left[\delta(s, a^*_{s,q}(u)) + cost(a^*_{s,q}(u), q) + \gamma v(a^*_{s,q}(u))\right] \\
&\quad - \mathbb{E}\left[\delta(s, a^*_{s,q}(u)) + cost(a^*_{s,q}(u), q) + \gamma u(a^*_{s,q}(u))\right] \\
&= \gamma\mathbb{E}\left[v(a^*_{s,q}(u)) - u(a^*_{s,q}(u))\right] \\
&\leq \gamma\mathbb{E}\left[\|v - u\|\right] = \gamma\|v - u\|.
\end{aligned}
$$

Thus we can conclude, $|\mathcal{M}v(s) - \mathcal{M}u(s)| \leq \gamma|v(s) - u(s)|$ for all configuration $s \in S$. From the definition of our norm, we can write

$$
\sup_{s \in S}|\mathcal{M}v(s) - \mathcal{M}u(s)| = \|\mathcal{M}v - \mathcal{M}u\| \leq \gamma\|v - u\|.
$$

This means that if $0 \leq \gamma < 1$, $\mathcal{M}$ is a contraction mapping. By [28, Proposition 3.10.2], there exists a unique v^* such that $\mathcal{M}v^* = v^*$, such that for an arbitrary v^0, the sequence v^n generated by $v^{n+1} = \mathcal{M}v^n$ converges to v^*. By the property of convergence of LSPI [18], $v^* = \tilde{V}$. From Proposition 1, the optimal cost-to-go function V^* also satisfies $\mathcal{M}V^* = V^*$. Hence $V^* = \tilde{V}$ and the property of convergence of LSPI is preserved in Algorithm 3.   □

5   Adaptive Database Tuning with Regularized Cost-Model Learning

In the results that we will present in Sect. 7.3, we will observe a higher variance of Algorithm 3 for index tuning than that of the state-of-art WFIT algorithm [39]. This high variance is caused mainly due to the absence of the cost model. As Algorithm 3 decides the policy depending on the estimated cost model, any error in the cost model causes instability in its outcome.

The process of cost-model estimation by accumulating information of incoming queries is analogous to approximating a function online from its incoming samples. Here, the function is the per-stage cost model $C\colon S \times S \times \tilde{Q} \to \mathbb{R}$. Here, $\tilde{Q}$ is the extended set of queries given by $Q \cup \{q(s, s') \mid s, s' \in S\}$. We obtain this $\tilde{Q}$ by considering configuration updates as special queries, as explained in Sect. 4.3. Now, the per-stage cost function can be defined as

$$
C(s_{t-1}, s_t, \hat{q}_t) = cost(s_{t-1}, q(s_{t-1}, s_t)) + cost(s_t, \hat{q}_t)
$$

This equation shows that if we consider changing the configuration from s to s' as executing a special query $q(s, s')$, approximating the function $cost\colon S \times \tilde{Q} \to \mathbb{R}$ in turn approximates the per-stage cost.

As explained in the previous section, we approximate $cost$ online using linear projection to the feature space of the observed state and query. At each step we obtain some vector ζ such that $cost(s,q) \approx \zeta^T \eta(s,q)$. Here, $\eta(s,q)$ is the feature vector corresponding to state s and query q. In order to obtain the optimal approximation, we initialize with an arbitrary ζ and then recursively improve our estimate of ζ using recursive least squares estimation (RLSE) algorithm [44]. But the issues with RLSE are:

i. It tries to minimize the square error per step

$$\hat{\epsilon}_t^2 = \left((\zeta^{t-1})^T \eta^t - cost(s_{t-1}, \hat{q}_t) \right)^2$$

which is highly sensitive to outliers. If RLSE faces some query or configuration update which is very different from the previously observed queries, the estimation of ζ can change drastically.

ii. The algorithm becomes unstable if the components of $\eta(s,q)$ are highly correlated. This may happen when the algorithm passes through a series of related queries.

As the reinforcement learning algorithm uses the estimated cost model to decide policies and to evaluate them, error or instability in the estimated cost model at any step affects its performance. Specifically, large deviations arise in the estimated cost model due to the queries which are far from the previously learned distribution. This costs the learning algorithm some time to adapt. It also affects the policy and evaluation of the present state and action. We thus propose to use a *regularized* cost-model estimator instead of RLSE, which is less sensitive to outliers and relatively stable, so as to improve the performance of Algorithm 3 and decrease its variance.

5.1   Regularized Cost-Model Estimator

In order to avoid the effect of outliers, we can penalize high variance of ζ by adding a regularization term with the squared prediction error of RLSE. Thus at time step t, the new estimator will try to find

$$\zeta^t = \arg\min_{\zeta} P^t \qquad \text{given} \qquad \hat{\epsilon}_t, \overline{B}^{t-1}, \zeta^{t-1}, \eta^t \qquad (5)$$

such that:

$$P^t := \hat{\epsilon}_t^2 + \lambda \|\zeta^{t-1}\|_2^2$$
$$= \left(\langle \zeta^{t-1}, \eta^t \rangle - cost(s_{t-1}, \hat{q}_t) \right)^2 + \lambda \langle \zeta^{t-1}, \zeta^{t-1} \rangle.$$

Here, $\lambda > 0$ is the regularization parameter. Square of L_2-norm, $\|\zeta\|_2^2$, is the regularization function. $\eta^t := \eta(s_{t-1}, \hat{q}_t)$ is the feature vector of state s_{t-1} and query $\hat{q}_t$. We call this squared error the *loss function* L_t defined at time t for a given choice of ζ^t. Thus,

$$L_t(\zeta^t) := \left(\langle \zeta^t, \eta^t \rangle - cost(s_{t-1}, \hat{q}_t) \right)^2.$$

The dual of this problem can be considered as picking up such an ζ^t inside an n-dimensional Euclidean ball $\mathbb{B}_\lambda^n$ of radius $s(\lambda)$ that minimizes the error $\hat{\epsilon}_t^2$. From an optimization point of view, we choose ζ^t inside $\mathbb{B}_\lambda^n :=$ $\{\zeta \mid \|\zeta\|_2^2 \le s(\lambda) \text{ and } \zeta \in \mathbb{R}^n\}$ rather than searching for it in the whole space $\mathbb{R}^n$. This estimator penalizes any drastic change in the cost model due to some outlier query. If some query tries to pull ζ^t out of $\mathbb{B}_\lambda^n$, this estimator regularizes ζ^t at the boundary. It also induces sparsity in the components of estimating vector ζ that eliminates the instability due to highly correlated queries.

Algorithm 4. Regularized cost-model estimation.

1: Initialize $\zeta^0 = 0$ and $R^0 = \varepsilon I$.
2: **for** t=1,2,3,... **do**
3:    $\hat{\epsilon}_t \leftarrow (\zeta^{t-1})^T \eta^t - cost(s_{t-1}, \hat{q}_t)$
4:    $\gamma^t \leftarrow \lambda + (\eta^t)^T R^{t-1} \eta^t$
5:    $R^t \leftarrow R^{t-1} - \frac{1}{\gamma^t}(R^{t-1}\eta^t(\eta^t)^T R^{t-1})$
6:    $\zeta^t \leftarrow \zeta^{t-1} - \frac{1}{\gamma^t}R^{t-1}\eta^t\hat{\epsilon}_t$
7:    **return** R^t, ζ^t
8: **end for**

The online penalized cost-model estimation algorithm obtained from this formulation is shown in Algorithm 4. Generally, the optimal values of ε and λ are decided using a cross-validation procedure. In Sect. 6.4, we are going to derive the optimal value of ε and a probable estimation for λ for the index tuning problem. This will decide optimal values of the hyper-parameters for a given set of workload with theoretical performance bounds.

5.2 Performance Bound

We can depict this online cost-model estimation task as a simple game between a decision maker and an adversary [35]. In database tuning, the decision maker is our cost-model estimating algorithm and the adversary is the workload providing an uncertain sequence of queries. Then, we can formulate the game as Algorithm 5.

Algorithm 5. Cost-model Estimation Game.

1: Initialize $\zeta^0 = 0$.
2: **for** t=1,2,3,..., T **do**
3:    Algorithm 4 picks $\zeta^t \in \mathbb{B}_\lambda^n$ according to Eq. (5)
4:    Adversary picks (η_t, c_t)
5:    Algorithm suffers loss $L_t(\zeta^t)$
6: **end for**

We can define the regret of this game after time step T as,

$$Reg_T := \sum_{t=1}^{T} L_t(\boldsymbol{\zeta}^t) - \sum_{t=1}^{T} L_t(\boldsymbol{\zeta}^{\text{OPT}}) \qquad (6)$$

where $\boldsymbol{\zeta}^{\text{OPT}}$ is the solution picked up by an offline expert that minimizes the cumulative loss after time step T. Reg_T is the difference between cumulative sum of errors up to time T obtained using Algorithm 4 and the optimal offline algorithm. This regret term captures deviation of the cost-model estimated by the Algorithm 5 from the computable optimal cost model.

As the loss function $L(\boldsymbol{\zeta})$ is the square of the error between estimated and actual values of cost at time t, it is a convex function over the set of $\boldsymbol{\zeta}$. According to the analysis given in [35], we can canonically describe our estimation model as a scheme to develop a Legendre potential function $\Phi(\boldsymbol{\zeta}^{\text{OPT}})$ with time t for the given workload, where the initial value of potential is given by:

$$\Phi_0(\boldsymbol{\zeta}^{\text{OPT}}) := \|\boldsymbol{\zeta}^{\text{OPT}}\|^2$$

and its value at time t is updated as

$$\Phi_t(\boldsymbol{\zeta}^{\text{OPT}}) := \Phi_{t-1}(\boldsymbol{\zeta}^{\text{OPT}}) + \frac{1}{\lambda} L_t(\boldsymbol{\zeta}^t).$$

Now, we can re-write Eq. (5) as:

$$\boldsymbol{\zeta}^t = \arg\min_{\boldsymbol{\zeta} \in \mathbb{B}_\lambda^n} \left[D_{\Phi_0}(\boldsymbol{\zeta}^{\text{OPT}}, \boldsymbol{\zeta}^{t-1}) + \frac{1}{\lambda} \left(\boldsymbol{\nabla} L_{t-1}(\boldsymbol{\zeta}_{t-1}) \right)^T \boldsymbol{\zeta}^{t-1} \right] \qquad (7)$$

Here, $D_{\Phi_0}(\boldsymbol{\zeta}^{\text{OPT}}, \boldsymbol{\zeta}^{t-1})$ is the Bregman divergence [26] between $\boldsymbol{\zeta}^{\text{OPT}}$ and $\boldsymbol{\zeta}^{t-1}$ along the potential field $\Phi_0(\boldsymbol{\zeta}^{\text{OPT}})$. This term in Eq. (7) inclines Algorithm 4 to choose such a $\boldsymbol{\zeta}$ which is nearest to optimal $\boldsymbol{\zeta}^{\text{OPT}}$ on the $\|\boldsymbol{\zeta}\|^2$ manifold. Also, $\boldsymbol{\nabla} L_{t-1}(\boldsymbol{\zeta}_{t-1})^T \boldsymbol{\zeta}^{t-1}$ is the change of the loss function in the direction of $\boldsymbol{\zeta}^{t-1}$. Minimization of this term is equivalent to selection of such a $\boldsymbol{\zeta}^t$ that minimizes the corresponding loss. Thus, the $\boldsymbol{\zeta}^t$ picked up by the Algorithm is the one that minimizes a linear combination of these two terms weighted by λ. From this formulation we can obtain the following lemma for the regret bound.

Lemma 1. *After time step T, the upper bound of the regret of Algorithm 4 can be given by*

$$Reg_T \leq \lambda \|\boldsymbol{\zeta}^{\text{OPT}}\|^2 + \frac{1}{\lambda} \sum_{t=1}^{T} \hat{\epsilon}_t^2 (\boldsymbol{\eta}^t)^T R^t \boldsymbol{\eta}^t. \qquad (8)$$

Proof. Applying Theorem 1 of [42] on Eq. (7) we get the inequalities,

$$Reg_T \leq \lambda \left[D_{\Phi_0}(\boldsymbol{\zeta}^{\text{OPT}}, \boldsymbol{\zeta}^0) - D_{\Phi_T}(\boldsymbol{\zeta}^{\text{OPT}}, \boldsymbol{\zeta}^{T+1}) + \sum_{t=1}^{T} D_{\Phi_t}(\boldsymbol{\zeta}^t, \boldsymbol{\zeta}^{t+1}) \right]$$

$$\leq \lambda \left[D_{\Phi_0}(\boldsymbol{\zeta}^{\text{OPT}}, \boldsymbol{\zeta}^0) + \sum_{t=1}^{T} D_{\Phi_t}(\boldsymbol{\zeta}^t, \boldsymbol{\zeta}^{t+1}) \right].$$

From the definition of the Legendre potential we get:

$$\Phi_t(\zeta) = \Phi_{t-1}(\zeta) + \frac{1}{\lambda}L_t(\zeta)$$

$$= \|\zeta\|^2 + \sum_{t=1}^{T}\left(\langle\zeta,\eta^t\rangle - cost(s_{t-1},\hat{q}_t)\right)^2$$

$$= \zeta^T\left(I + \frac{1}{\lambda}\sum_{t=1}^{T}\eta^t(\eta^t)^T\right)\zeta - \zeta^T\left(\frac{1}{\lambda}\sum_{t=1}^{T}cost(s_{t-1},\hat{q}_t)\eta^t\right) + \sum_{t=1}^{T}cost(s_{t-1},\hat{q}_t)^2$$

$$= \zeta^T(R^T)^{-1}\zeta - \zeta^T b_T + C_T$$

where $b_T = \sum_{t=1}^{T}c^t\eta^t$ and $C_T = \sum_{t=1}^{T}cost(s_{t-1},\hat{q}_t))^2$. Thus, the dual of the potential can be given by

$$\Phi_t^*(\zeta) = \zeta^T R^T\zeta - 2\zeta^T R^T b_T + (b_T)^T R^T b_T$$

Now, from the definition of Φ_0 and properties of Bregman divergence,

$$D_{\Phi_0}(\zeta^{OPT},\zeta^0) = D_{\|\zeta^{OPT}\|^2}(\zeta^{OPT},\zeta^0)$$
$$= \|\zeta^{OPT}\|^2$$

and

$$D_{\Phi_t}(\zeta^t,\zeta^{t+1}) = D_{\Phi_t^*}\left(\nabla\Phi_t(\zeta^{t+1}),\nabla\Phi_t(\zeta^t)\right)$$
$$= D_{\Phi_t^*}\left(0,\nabla\Phi_t(\zeta^t)\right)$$
$$= D_{\Phi_t^*}\left(0,\frac{1}{\lambda}\nabla L_t(\zeta^t)\right)$$
$$= \frac{1}{\lambda^2}\left(\nabla L_t(\zeta^t)\right)^T R^t\left(\nabla L_t(\zeta^t)\right)$$
$$= \frac{1}{\lambda^2}\left(\langle\zeta,\eta^t\rangle - cost(s_{t-1},\hat{q}_t)\right)^2(\eta^t)^T R^t\eta^t$$
$$= \frac{1}{\lambda^2}\hat{\epsilon}_t^{\,2}(\eta^t)^T R^t\eta^t.$$

By replacing these results in the aforementioned inequality we get:

$$Reg_T \le \lambda\|\zeta^{OPT}\|^2 + \frac{1}{\lambda}\sum_{t=1}^{T}\hat{\epsilon}_t^{\,2}(\eta^t)^T R^t\eta^t.$$

$\square$

Lemma 2. *If $R^0 \in \mathbb{R}^{n\times n}$ and invertible,*

$$(\eta^t)^T R^t\eta^t = 1 - \frac{\det(R^t)}{\det(R^{t-1})} \quad \forall t = 1,2,\ldots,T \tag{9}$$

Proof. From [19], we get if there exists an invertible matrix $B \in \mathbb{R}^{n \times n}$ such that $A = B + \boldsymbol{x}\boldsymbol{x}^T$, where $\boldsymbol{x} \in \mathbb{R}^n$, then

$$\boldsymbol{x}^T A^{-1} \boldsymbol{x} = 1 - \frac{\det(B)}{\det(A)} \tag{10}$$

As, per Algorithm 4, $R^0 = \varepsilon I$, it is invertible. Since $(R^t)^{-1} = (R^{t-1})^{-1} + \boldsymbol{\eta}^t (\boldsymbol{\eta}^t)^T$, by the Sherman–Morrison formula, all R^t's are invertible for $t \geq 0$. Thus, simply replacing A by $(R^t)^{-1}$ and B by $(R^{t-1})^{-1}$ in Eq. (10), we obtain

$$(\boldsymbol{\eta}^t)^T R^t \boldsymbol{\eta}^t = 1 - \frac{\det((R^{t-1})^{-1})}{\det((R^t)^{-1})} = 1 - \frac{\det(R^t)}{\det(R^{t-1})}$$

since, $\det((R^t)^{-1}) = \frac{1}{\det(R^t)}$.   □

Using Lemmas 1 and 2, we finally derive the regret bound for the regularized cost-model estimator in the following theorem.

Theorem 1. *If we consider the error as a bounded function such that $0 \leq \hat{\epsilon}_t^2 \leq E_{\max}$ and $\|\boldsymbol{\eta}^t\|_\infty \leq \delta$,*

$$Reg_T \leq \lambda \|\boldsymbol{\zeta}^{\mathrm{OPT}}\|^2 + \frac{E_{\max}}{\lambda} \left[n \ln\left(1 + \frac{\varepsilon \delta^2 T}{n}\right) - (n-1)\ln(\varepsilon) \right] \tag{11}$$

where $R^0 = \varepsilon I$.

Proof. Let us assume the squared error has an upper bound $E_{\max}$ for a given workload. Under this assumption, we get from Eqs. (8) and (9),

$$Reg_T \leq \lambda \|\boldsymbol{\zeta}^{\mathrm{OPT}}\|^2 + \frac{E_{\max}}{\lambda} \sum_{t=1}^{T} \left(1 - \frac{\det(R^t)}{\det(R^{t-1})}\right)$$

$$\leq \lambda \|\boldsymbol{\zeta}^{\mathrm{OPT}}\|^2 - \frac{E_{\max}}{\lambda} \sum_{t=1}^{T} \ln\left(\frac{\det(R^t)}{\det(R^{t-1})}\right)$$

$$= \lambda \|\boldsymbol{\zeta}^{\mathrm{OPT}}\|^2 + \frac{E_{\max}}{\lambda} \ln\left(\frac{\det(R^0)}{\det(R^T)}\right)$$

$$= \lambda \|\boldsymbol{\zeta}^{\mathrm{OPT}}\|^2 + \frac{E_{\max}}{\lambda} \left[\ln(\varepsilon) - \ln(\det(R^T))\right]$$

$$= \lambda \|\boldsymbol{\zeta}^{\mathrm{OPT}}\|^2 + \frac{E_{\max}}{\lambda} \left[\ln(\varepsilon) + \ln\left(\det\left(\frac{1}{\varepsilon}I + \sum_{t=1}^{T} \boldsymbol{\eta}^t (\boldsymbol{\eta}^t)^T\right)\right)\right]$$

$$= \lambda \|\boldsymbol{\zeta}^{\mathrm{OPT}}\|^2 + \frac{E_{\max}}{\lambda} \left[\sum_{k=1}^{n} \ln(1 + \varepsilon \lambda_k) - (n-1)\ln(\varepsilon)\right].$$

Because

$$\det\left(\frac{1}{\varepsilon}I + \sum_{t=1}^{T} \boldsymbol{\eta}^t (\boldsymbol{\eta}^t)^T\right) = \varepsilon^{-n} \det\left(I + \varepsilon \sum_{t=1}^{T} \boldsymbol{\eta}^t (\boldsymbol{\eta}^t)^T\right) = \varepsilon^{-n} \prod_{k=1}^{n} (1 + \varepsilon \lambda_k)$$

where $\lambda_1, \ldots, \lambda_n$ are eigenvalues of the matrix $\sum_{t=1}^{T} \boldsymbol{\eta}^t (\boldsymbol{\eta}^t)^T$. As the eigenvalues of $\sum_{t=1}^{T} \boldsymbol{\eta}^t (\boldsymbol{\eta}^t)^T$ are equal to the eigenvalues of its Gram matrix $G_{ij} = (\boldsymbol{\eta}^i)^T \boldsymbol{\eta}^j$, we can write

$$\sum_{k=1}^{n} \lambda_k = Trace(G) = \sum_{t=1}^{T} (\boldsymbol{\eta}^t)^T \boldsymbol{\eta}^t \leq \delta^2 T$$

where $\|\boldsymbol{\eta}^t\|_\infty \leq \delta$, that is, the maximum value of any component of $\boldsymbol{\eta}$ is bounded by δ. In the above inequality, the equality holds if and only if $\lambda_1 = \lambda_2 = \ldots = \lambda_n = \frac{\delta^2 T}{n}$. By applying this condition, we get the regret bound as

$$Reg_T \leq \lambda \|\boldsymbol{\zeta}^{\text{OPT}}\|^2 + \frac{E_{\max}}{\lambda} \left[n \ln \left(1 + \frac{\varepsilon \delta^2 T}{n} \right) - (n-1)\ln(\varepsilon) \right].$$

$\square$

This theorem shows that our estimation of the cost model using Algorithm 4 is always upper bounded by a constant value depending on the optimal solution added with a term that increases with time logarithmically. This shows that the regret, which is the cumulative deviation of the cost model computed by Algorithm 4 with respect to the optimal one, increases very slowly with time. That means the error of estimation in each and every time step is considerably small.

6   Case Study: Index Tuning

In this section we present *COREIL* (for *Cost-model Oblivious REInforcement Learning algorithm*) and its regularized version, *rCOREIL*. COREIL and rCOR-EIL instantiate Algorithm 3 taking as cost-model estimators Algorithms 2 and 4 respectively. Both of them tune the configurations differing in their secondary indexes and handle the configuration changes corresponding to the creation and deletion of indexes. COREIL uses reinforcement learning approach to solve the index tuning problem *on-the fly*. It projects index tuning as an MDP and applies Algorithm 3 to solve it. On the other hand, rCOREIL uses the regularized cost-model estimator, described in Sect. 5.1; rCOREIL's regularized estimator affords it to leverage the fact that if we serve the learning algorithm with a better cost-model to evaluate its policy better, it will perform better. In this section, we also define the feature mappings ϕ and $\boldsymbol{\eta}$ for both COREIL and rCOREIL. They are used to approximate the cost-to-go function V and the *cost* function respectively. At the end of this section we prove tighter performance bounds for Algorithm 4 in case of index tuning. We also derive optimal values of the parameters λ and ε for a given workload.

6.1   Reducing the Search Space

Let I be the set of indexes that can be created. Each configuration $s \in S$ is an element of the power set 2^I. For example, 7 attributes in a schema of R yield a total of 13699 indexes and a total of 2^{13699} possible configurations. Such a large search space invalidates a naive brute-force search for the optimal policy.

For any query $\hat{q}$, let $r(\hat{q})$ be a function that returns a set of recommended indexes. This function may be already provided by the database system (e.g., as with IBM DB2), or it can be implemented externally [1]. Let $d(\hat{q}) \subseteq I$ be the set of indexes being modified (update, insertion or deletion) by $\hat{q}$. We can define the reduced search space as

$$S_{s,\hat{q}} = \{s' \in S \mid (s - d(\hat{q})) \subseteq s' \subseteq (s \cup r(\hat{q}))\}. \tag{12}$$

Deleting indexes in $d(\hat{q})$ will reduce the index maintenance overhead and creating indexes in $r(q)$ will reduce the query execution cost. Note that the definition of $S_{s,\hat{q}}$ here is a subset of the one defined in Sect. 4.2 which deals with the general configurations.

Note that for tree-structured indexes (e.g., B$^+$-tree), we could further consider the *prefix closure* of indexes for optimization. For any configuration $s \in 2^I$, define the prefix closure of s as

$$\langle s \rangle = \{i \in I \mid i \text{ is a prefix of an index } j \text{ for some } j \in s\}. \tag{13}$$

Thus in Eq. (12), we use $\langle r(\hat{q}) \rangle$ to replace $r(\hat{q})$ for better approximation. The intuition is that in case of $i \notin s$ but $i \subseteq \langle s \rangle$ we can leverage the prefix index to answer the query.

6.2   Defining the Feature Mapping ϕ

Let V be the cost-to-go function following a policy. As mentioned earlier, Algorithm 3 relies on a proper feature mapping ϕ that approximates the cost-to-go function as $V(s) \approx \theta^T \phi(s)$ for some vector θ. The challenge lies in how to define ϕ under the scenario of index tuning. Both in COREIL and rCOREIL, we define it as

$$\phi_{s'}(s) := \begin{cases} 1, & \text{if } s' \subseteq s \\ -1, & \text{otherwise} \end{cases}$$

for each $s, s' \in S$. Let $\phi = (\phi_{s'})_{s' \in S}$. Note that $\phi_\emptyset$ is an intercept term since $\phi_\emptyset(s) = 1$ for all $s \in S$. The following proposition shows the effectiveness of ϕ for capturing the values of the cost-to-go function V.

Proposition 3. *There exists a unique* $\theta = (\theta_{s'})_{s' \in S}$ *which approximates the value function as*

$$V(s) = \sum_{s' \in S} \theta_{s'} \phi_{s'}(s) = \theta^T \phi(s). \tag{14}$$

Proof. Suppose $S = \{s^1, s^2, \dots, s^{|S|}\}$. Note that we use superscripts to denote the ordering of elements in S.

Let $V = (V(s))_{s \in S}^T$ and M be a $|S| \times |S|$ matrix such that

$$M_{i,j} = \phi_{s^j}(s^i).$$

Let θ be a $|S|$-dimension column vector such that $M\theta = V$. If M is invertible then $\theta = M^{-1}V$ and thus Eq. (14) holds.

We now show that M is invertible. Let ψ be a $|S| \times |S|$ matrix such that

$$\psi_{i,j} = M_{i,j} + 1.$$

We claim that ψ is invertible and its inverse is the matrix τ such that

$$\tau_{i,j} = (-1)^{|s^i|-|s^j|}\psi_{i,j}.$$

To see this, consider

$$(\tau\psi)_{i,j} = \sum_{1 \le k \le |S|} (-1)^{|s^i|-|s^k|}\psi_{i,k}\psi_{k,j}$$

$$= \sum_{s_j \subseteq s_k \subseteq s_i} (-1)^{|s^i|-|s^k|}.$$

Therefore $(\tau\psi)_{i,j} = 1$ if and only if $i = j$. By the Sherman-Morrison formula, M is also invertible.

However, for any configuration s, $\theta(s)$ is a $|2^I|$-dimensional vector. In order to reduce the dimensionality, the cost-to-go function can be approximated by $V(s) \approx \sum_{s' \in S, |s'| \le N} \theta_{s'}\phi_{s'}(s)$ for some integer N. Here we assume that the collaborative benefit among indexes could be negligible if the number of indexes exceeds N. In particular when $N = 1$, we have

$$V(s) \approx \theta_0 + \sum_{i \in I} \theta_i\phi_i(s). \tag{15}$$

where we ignore all the collaborative benefits among indexes in a configuration. This is reasonable since any index in a database management system is often of individual contribution for answering queries [31]. Therefore, we derive ϕ from Eq. (15) as $\phi(s) = (1, (\phi_i(s))_{i \in I}^T)^T$. By using this feature mapping ϕ, both COREIL and rCOREIL approximate the cost-to-go function $V(s) \approx \theta^T\phi(s)$ for some vector θ.

6.3 Defining the Feature Mapping η

A good feature mapping for approximating functions δ and *cost* must take into account both the benefit from the current configuration and the maintenance overhead of the configuration.

To capture the difference between the index set recommended by the database system and that of the current configuration, we define a function $\beta(s, \hat{q}) = (1, (\beta_i(s, \hat{q}))_{i \in I}^T)^T$, where

$$\beta_i(s, \hat{q}) := \begin{cases} 0, & i \notin r(\hat{q}) \\ 1, & i \in r(\hat{q}) \text{ and } i \in s \\ -1, & i \in r(\hat{q}) \text{ and } i \notin s. \end{cases}$$

If the execution of query $\hat{q}$ cannot benefit from index i then $\beta_i(s, \hat{q})$ always equals zero; otherwise, $\beta_i(s, \hat{q})$ equals 1 or -1 depending on whether s contains i or not. For tree-structured indexes, we could further consider the prefix closure of indexes as defined in Eq. (13) for optimization.

On the other hand, to capture whether a query (update, insertion or deletion) modifies any index in the current configuration, we define a function $\alpha(s, \hat{q}) = (\alpha_i(s, \hat{q}))_{i \in I}$ where

$$\alpha_i(s, \hat{q}) = \begin{cases} 1, & \text{if } i \in s \text{ and } \hat{q} \text{ modify } i \\ 0, & \text{otherwise.} \end{cases}$$

Note that if $\hat{q}$ is a selection query, α trivially returns $\mathbf{0}$.

By combining β and α, we get the feature mapping $\eta = (\beta^T, \alpha^T)^T$ used in both of the algorithms. It can be used to approximate the functions δ and $cost$ as described in Sect. 4.3.

6.4   Performance Bounds for Regularized COREIL

rCOREIL applies Algorithm 4 for cost-model estimation, while COREIL uses RLSE for this. If we follow Algorithm 3, on line 13 rCOREIL calls the regularized cost-model estimator with arguments $\hat{\epsilon}^t, R^{t-1}, \zeta^{t-1}, \eta^t$ instead of RLSE. Following Theorem 1 and the construction of the feature map in Sect. 6.3, Proposition 4 gives a tighter regret bound for the cost-model estimation of rCOREIL.

Proposition 4. *If we consider the error as a bounded function such that* $0 \leq \hat{\epsilon}_t^2 \leq E_{\max}$:

$$Reg_T^{rCOREIL} \leq \lambda \|\zeta^{OPT}\|^2 + \frac{E_{\max}}{\lambda} [2n\ln T - n\ln n] \tag{16}$$

and the optimal value for ε is given by:

$$\varepsilon^* = \frac{n^2 - n}{T}.$$

Proof. From Sect. 6.3, $\|\eta^t\|_\infty \leq 1$. Equation (11) transforms into

$$Reg_T^{rCOREIL} \leq \lambda \|\zeta^{OPT}\|^2 + \frac{E_{\max}}{\lambda} \left[n\ln\left(1 + \frac{\epsilon T}{n}\right) - (n-1)\ln(\varepsilon) \right].$$

Now, we determine the optimal value of ε by minimizing the RHS of above inequality as this will impose tighter limit on the bound. Thus,

$$\left[\frac{\partial(\mathrm{RHS})}{\partial\varepsilon}\right]_{\varepsilon^*=0} = 0.$$

By solving this, we get $\varepsilon^* = \frac{n^2-n}{T}$. Substituting this value in the previous inequality gives us the regret bound for regularized COREIL algorithm as

$$Reg_T^{rCOREIL} \leq \lambda\|\zeta^{OPT}\|^2 + \frac{E_{\max}}{\lambda}\left[2n\ln(T) - n\ln(n)\right].$$

$\square$

Similarly, we can also find out the optimal value of λ that will make the upper bound tightest.

Corollary 1. *If the value of optimal solution ζ^{OPT} can be predicted beforehand, the optimal value of λ is given by*

$$\lambda^* = \frac{E_{\max}}{\|\zeta^{OPT}\|^2}\left[2n\ln(T) - n\ln(n)\right]$$

where the stopping time T is given.

Proof. As an optimal λ will minimize the RHS of Eq. (16), we get it by setting the partial derivative of the RHS with respect to λ as zero. This simply gives us, $\lambda^* = \frac{E_{\max}}{\|\zeta^{OPT}\|^2}\left[2n\ln(T) - n\ln(n)\right]$.

Substituting the optimal value of λ in Eq. (16) for a given T and ζ^{OPT}, we get

$$Reg_T^{rCOREIL} \leq \|\zeta^{OPT}\|^2 + E_{\max}\left[2n\ln(T) - n\ln(n)\right].$$

For large n and comparatively smaller T, $[2n\ln(T) - n\ln(n)]$ is a negative number that makes the plausible error in cost-model estimation much smaller than even the magnitude of the optimal ζ vector. This shows the guarantee on the quality of the cost-model estimated by rCOREIL once the parameters are properly set.

7   Performance Evaluation

In this section, we present an empirical evaluation of COREIL and rCOREIL through two sets of experiments. In the first set of experiments, we implement a prototype of COREIL in Java. We compare its performance with that of the state-of-the-art WFIT algorithm [39] (briefly described in Sect. 7.2). In the results, we can see that COREIL shows competitive performance with WFIT but has higher variance. This validates the efficiency of the reinforcement learning approach to solve the index tuning problem *on the fly*. This shows that, even without any assumption of a pre-determined cost model, it is possible to perform at the level of the state-of-the-art.

In the second set of experiments, we evaluate the performance of rCOR-EIL with respect to COREIL. The results show enhancements in performance by rCOREIL as reasoned in Sect. 5. This validates the claim in Sect. 5 that the higher variance of COREIL is due to suboptimal use of the RLSE algorithm. It also establishes the fact that if we serve the learning algorithm with an enhanced estimation of cost-model, it improves the performance substantially. In these experiments, we also check the sensitivity of rCOREIL with respect to the parameter λ and cross-validate the optimal value for the given workload.

7.1 Dataset and Workload

The dataset and workload is conforming to the TPC-C specification [30] and generated by the OLTP-Bench tool [15]. The 5 types of transactions in TPC-C are distributed as NewOrder 45%, Payment 43%, OrderStatus 4%, Delivery 4% and StockLevel 4%. Each of these transactions are associated with $3 \sim 5$ SQL statements (query/update). The scale factor used throughout the experiments is 2. We do not leverage any repetition or periodicity of the workload in our approach; still for robustness there may be up to 10 % of repetition of queries. Note that [39] additionally uses the dataset NREF in its experiments. However, this dataset and workload are not publicly available.

7.2 WFIT: Brief Description

WFIT is proposed in [39] as a method of semi-automatic index tuning. This algorithm keeps the database administrator "in the loop" by generating recommendations. These recommendations are generated through a feedback loop originating from the administrator's preferences. This process is based on the Work Function Algorithm [8]. In order to determine the change of configuration, WFIT considers all the queries observed in the past. Then it solves a deterministic problem of minimizing the total processing cost. However, while doing so, it assumes the existence of a pre-determined cost model served by the database system or administrator. Due to use of a pre-defined cost model for all the datasets and workloads it faces the problems discussed in the Introduction. Results documented in the following sections will show the importance of a reinforcement learning approach to make the process generic and cost-model oblivious.

7.3 COREIL: Experiments and Results

Experimental Set-Up. We conduct all the experiments on a server running IBM DB2 10.5. The server is equipped with Intel i7-2600 Quad-Core @ 3.40 GHz and 4 GB RAM. We measure wall-clock times for execution of all components. Specially, for execution of workload queries or index creating/dropping, we measure the response time of processing corresponding SQL statement in DB2. Additionally, WFIT uses the what-if optimizer of DB2 to evaluate the cost. In this setup, each query is executed only once and all the queries were generated from one execution history.

Efficiency. Figure 1 shows the total cost of processing TPC-C queries for online index tuning of COREIL and WFIT. Total cost consists of the overhead of corresponding tuning algorithm, cost of configuration change and that of query execution. Results show that, after convergence, COREIL has lower processing cost most of the time. But COREIL converges slower than WFIT, which is expected since it does not rely on the what-if optimizer to guide the index

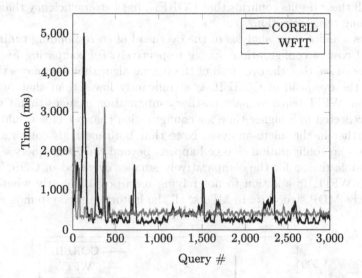

Fig. 1. Evolution of the efficiency (total time per query) of the two systems from the beginning of the workload (smoothed by averaging over a moving window of size 20)

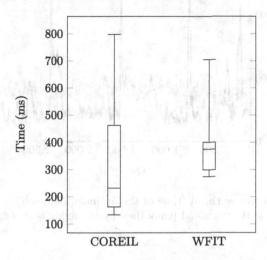

Fig. 2. Box chart of the efficiency (total time per query) of the two systems. We show in both cases the 9th and 91th percentiles (whiskers), first and third quartiles (box) and median (horizontal rule).

creations.[1] With respect to the whole execution set, the average processing cost of COREIL (451 ms) is competitive to that of WFIT (452 ms). However, if we calculate the average processing cost of the 500th query forwards, the average performance of COREIL (357 ms) outperforms that of WFIT (423 ms). To obtain further insight from these data, we study the distribution of the processing time per query, as shown in Fig. 2. As can be seen, although COREIL exhibits larger variance in the processing cost, its median is significantly lower that that of WFIT. All these results confirms that COREIL has better efficiency than WFIT under a long term execution.

Figures 3 and 4 show analysis of the overhead of corresponding tuning algorithm and cost of configuration change respectively. By comparing Fig. 1 with Fig. 3, we can see that the overhead of the tuning algorithm dominates the total cost and the overhead of COREIL is significantly lower than that of WFIT. In addition, WFIT tends to make costlier configuration changes than COREIL, which is reflected in a higher time for configuration change. This would be discussed further in the micro-analysis. Note that both methods converge rather quickly and no configuration change happens beyond the 700th query.

A possible reason for the comparatively smaller overhead of COREIL with respect to WFIT, in addition to not relying on a possibly costly what-if optimizer, is the MDP structure. In MDPs, all the history of the system is assumed

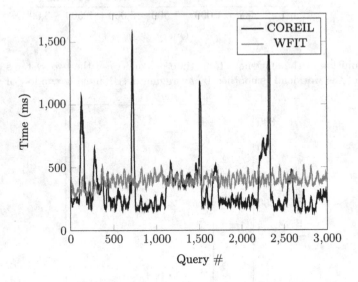

Fig. 3. Evolution of the overhead (time of the optimization itself) of the two systems from the beginning of the workload (smoothed by averaging over a moving window of size 20)

---

[1] By convergence we mean the first stable patch in Fig. 1 after the series of high spikes, around the 500th query. The convergence point is qualitatively chosen by observing characteristics of the curve.

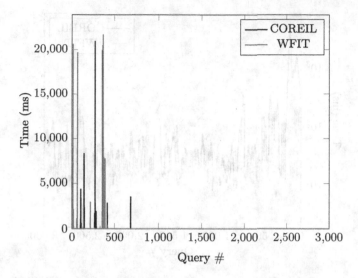

Fig. 4. Evolution of the time taken by configuration change (index creation and destruction) of the two systems from the beginning of the workload; no configuration change happens past query #700. All values except the vertical lines shown in the figure are zero.

to be summarized in the present state and the cost-function. Thus, COREIL has to do less book-keeping than WFIT.

Effectiveness. To verify the effectiveness of indexes created by the tuning algorithms, we extract the cost of query execution from the total cost. Figure 5 (note the logarithmic y-axis) indicates that the set of indexes created by COREIL shows competitive effectiveness with that created by WFIT, though WFIT is more effective in general and exhibits less variance after convergence. Again, this is to be expected since COREIL does not have access to any cost model for the queries. As previously noted, the total running time is lower for COREIL than WFIT, as overhead rather than query execution dominates running time for both systems.

We have also performed a micro-analysis to check whether the indexes created by the algorithms are reasonable. We observe that WFIT creates more indexes with longer compound attributes, whereas COREIL is more parsimonious in creating indexes. For instance, WFIT creates a 14-attribute index as shown below.

```
[S_W_ID, S_I_ID, S_DIST_10, S_DIST_09, S_DIST_08, S_DIST_07,
 S_DIST_06, S_DIST_05, S_DIST_04, S_DIST_03, S_DIST_02,
 S_DIST_01, S_DATA, S_QUANTITY]
```

The reason of WFIT creating such a complex index is probably due to multiple queries with the following pattern.

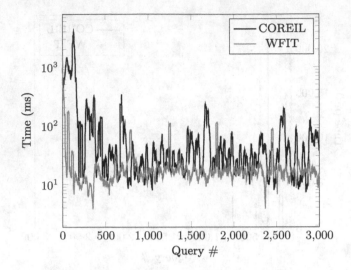

Fig. 5. Evolution of the effectiveness (query execution time in the DBMS alone) of the two systems from the beginning of the workload (smoothed by averaging over a moving window of size 20); logarithmic y-axis

```
SELECT S_QUANTITY, S_DATA, S_DIST_01, S_DIST_02, S_DIST_03,
       S_DIST_04, S_DIST_05, S_DIST_06, S_DIST_07, S_DIST_08,
       S_DIST_09, S_DIST_10
FROM STOCK
WHERE S_I_ID = 69082 AND S_W_ID = 1;
```

In contrast, COREIL tends to create shorter compound-attribute indexes. For example, COREIL created an index [S_I_ID, S_W_ID] which is definitely beneficial to answer the query above and is competitive in performance compared with the one created by WFIT.

7.4   rCOREIL: Experiments and Results

Experimental Set-Up. We run COREIL and rCOREIL, with a set of λ values $300, 350, 400, 450,$ and 500. The previous set of experiments have already established competitive performance of COREIL with WFIT. In this set we evaluate the basic idea of rCOREIL: providing regularized estimation of cost-model enhances the performance of COREIL and also stabilizes it. We conduct all the experiments on a server running IBM DB2 10.5 with scale factor and time measure, mentioned in the previous set of experiments. But here the server is installed on a 64 bit Windows virtual box with dual-core 2-GB hard disk. It operates in an Ubuntu machine with Intel i7-2600 Quad-Core @ 3.40 GHz and 4 GB RAM. This eventually makes both version of algorithms slower in comparison to the previous physical machine installation.

Efficiency. As the offline optimal outcome for this workload is unavailable beforehand, we set an expected range of λ as $[300, 600]$ depending on the other parameters like the number of queries and the size of state space. Figure 6 shows efficiency of COREIL and rCOREIL with different values of λ. As promised by Algorithm 4, variations of rCOREIL are always showing lesser median and

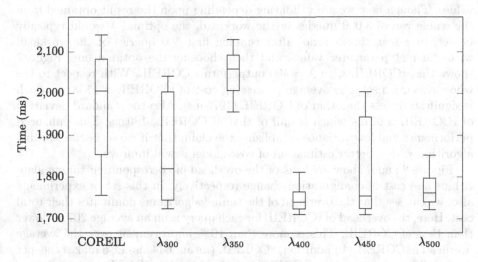

Fig. 6. Box chart of the efficiency (total time per query) of COREIL and its improved version with different values of λ. We show in both cases the 9th and 91st percentile (whiskers), first and third quartiles (box) and median (horizontal rule).

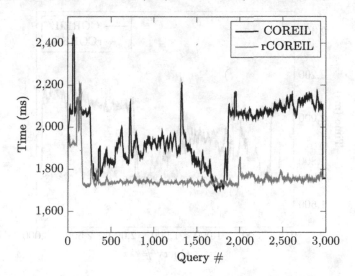

Fig. 7. Evolution of the efficiency (total time per query) of COREIL and rCOREIL with $\lambda = 400$ from the beginning of the workload (smoothed by averaging over a moving window of size 20)

variance of total cost. We can also observe from the boxplot, the efficiency is maximum as well as the variance is minimum for $\lambda = 400$. As efficiency is the final measure that controls runtime performance of the algorithm, we have considered this as optimal value of λ for further analysis. This process is analogous to cross-validation of parameter λ, where the proved bounds help us to set a range of values for searching it instead of going through an arbitrary large range of values. Though here we are validating depending upon the result obtained from the whole run of 3,000 queries in the workload, the optimal λ would typically be set, in a realistic scenario, after running first 500 queries of the workload with different parameter values and then choosing the optimal one. Figure 7 shows that rCOREIL with $\lambda = 400$ outperforms COREIL. With respect to the whole execution set, the average processing cost of rCOREIL is 1758 ms which is significantly less than that of COREIL (1975 ms). Also the standard deviation of rCOREIL is 90 ms which is half of that of COREIL, 180 ms. This enhanced performance and low variance establishes the claim that if we serve the learning algorithm with a better estimation of cost-model it will improve.

Figures 8 and 9 show analysis of the overhead of corresponding tuning algorithms and cost of configuration change respectively. In this set of experiments also, we can see that the overhead of the tuning algorithms dominates their total cost. Here, the overhead of rCOREIL for each query is on an average 207 ms lower than that of COREIL. This is more than 10 % improvement over the average overhead of COREIL. In addition, rCOREIL (mean: 644 ms) also makes cheaper configuration changes than COREIL (mean: 858 ms). rCOREIL also converges faster than COREIL as the last configuration update made by rCOREIL occurs

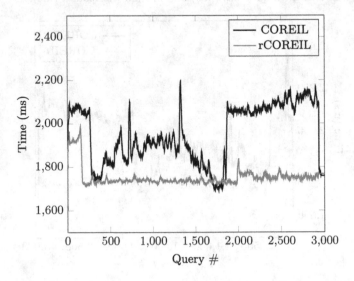

Fig. 8. Evolution of the overhead (time of the optimization itself) of COREIL and rCOREIL with $\lambda = 400$ from the beginning of the workload (smoothed by averaging over a moving window of size 20)

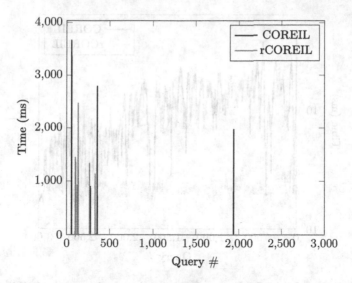

Fig. 9. Evolution of the time taken by configuration change (index creation and destruction) of COREIL and rCOREIL with $\lambda = 400$ from the beginning of the workload; no configuration change happens past query #2000. All values except the vertical lines shown in the figure are zero.

at the 335th query but the last two updates for COREIL occur at the 358th and 1940th queries respectively. If we look closely, the 358th and 1940th queries in this particular experiment are:

```
SELECT COUNT(DISTINCT (S_I_ID)) AS STOCK_COUNT
FROM ORDER_LINE, STOCK
WHERE OL_W_ID = 2 AND OL_D_ID = 10 AND OL_O_ID < 3509
     AND OL_O_ID >= 3509 - 20 AND S_W_ID = 2
     AND S_I_ID = OL_I_ID AND S_QUANTITY < 20;
```

and

```
SELECT COUNT(DISTINCT (S_I_ID)) AS STOCK_COUNT
FROM ORDER_LINE, STOCK
WHERE OL_W_ID = 1 AND OL_D_ID = 8 AND OL_O_ID < 3438
     AND OL_O_ID >= 3438 - 20 AND S_W_ID = 1
     AND S_I_ID = OL_I_ID AND S_QUANTITY < 11;
```

In reaction to this, COREIL creates indexes [ORDER_LINE.OL_D_ID,ORDER_LINE. +OL_W_ID] and [STOCK.S_W_ID, STOCK.S_QUANTITY] respectively. It turns out that such indexes are not of much use for most other queries (only 6 out of 3000 queries benefit of one of these indexes). COREIL makes configuration updates to tune the indexes for such queries, while the regularized cost model of rCOREIL does not make configuration updates due to rare and complex events, because it regularizes any big change due to such an outlier. Instead, rCOREIL has a

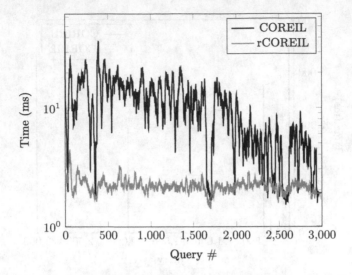

Fig. 10. Evolution of the effectiveness (query execution time in the DBMS alone) of COREIL and rCOREIL with $\lambda = 400$ from the beginning of the workload (smoothed by averaging over a moving window of size 20); logarithmic y-axis

slightly higher the overhead to find out the optimal indexes. For example, in the window consisting of 10 queries after the 359th query average overhead of rCOREIL increases from 1724 ms to 1748 ms.

Effectiveness. Like Sect. 7.3, here also we extract the cost of query execution to verify the effectiveness of indexes created by the tuning algorithms. Figure 10 indicates that the set of indexes created by rCOREIL are significantly more effective than those created by COREIL. We can see the average query execution time of rCOREIL is less than that of COREIL almost by a factor of 10.

At a micro-analysis level, we observe rCOREIL creates only one index with two combined attributes, all other indexes being single-attribute. On the other hand, COREIL creates only one index with a single attribute whereas all other indexes have two attributes. This observation shows that though COREIL creates parsimonious and efficient indexes, rCOREIL shows even better specificity and effectiveness in doing so.

7.5   Analysis of Cost Estimator

In order to examine the quality of the three cost estimators used by WFIT, COREIL, and rCOREIL to predict the actual cost of query executions or configuration updates, we observe the actual execution time, the estimated cost, and that returned by the what-if optimizer during every run of experiments for COREIL and rCOREIL, respectively. The scatter plot of Fig. 11 shows that the what-if cost has significantly less correlation (0.013) with the actual execution

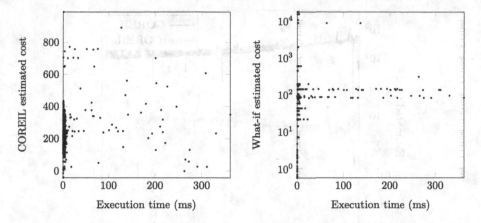

Fig. 11. Scatter plot of the estimated cost by COREIL and the what-if optimizer vs execution time. Left shows correlation between cost estimated by COREIL and actual execution time (in ms). Right shows (on a log y-axis) correlation between the cost estimated by the what-if optimizer and the actual execution time (in ms) in the same run.

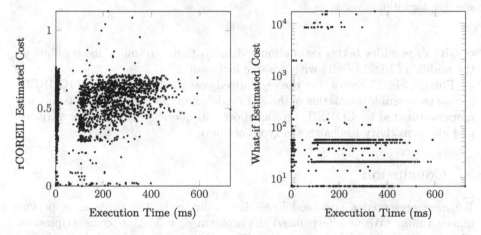

Fig. 12. Scatter plot of the estimated cost by rCOREIL and the what-if optimizer vs execution time. Left shows correlation between cost estimated by rCOREIL and actual execution time (in ms). Right shows (on a log y-axis) correlation between the cost estimated by the what-if optimizer and the actual execution time (in ms) in the same run.

time than COREIL (0.1539) Again, the scatter plot of Fig. 12 shows the regularized cost estimated by rCOREIL has significantly higher positive correlation (0.1558) than that predicted by the what-if optimizer. This proves that the execution time estimated by COREIL and rCOREIL are significantly more reliable than the ones estimated by what-if optimizer. It can also been observed that

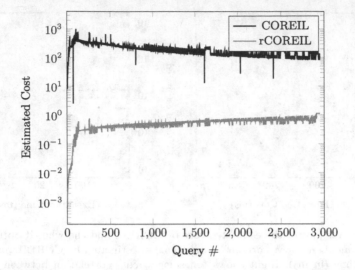

Fig. 13. Evolution of the estimated costs of COREIL and rCOREIL with $\lambda = 400$ from the beginning of the workload (smoothed by averaging over a moving window of size 20); logarithmic y-axis

rCOREIL provides better estimations: visually, there are many more points at the middle of Fig. 12 (left) with positive inclination.

Finally, Fig. 13 shows that the regularized cost model estimator of rCOREIL gives a more stable estimation of the cost model than that of COREIL, as the cost model estimated by COREIL (averaged over 20 queries) shows higher variance and also sensitivity to changes in types of queries.

8   Conclusion

We have presented a cost-model oblivious solution to the problem of performance tuning. We first formalized the problem as a Markov decision process. Then we devised and presented a solution, which addresses both issues of the curse of dimensionality and of over-fitting. We instantiated the problem to the case of index tuning. For this case, we implemented and evaluated the COR-EIL and rCOREIL algorithms, with and without regularization, respectively. Experiments show competitive performance with respect to the state-of-the-art WFIT algorithm, despite our approach being cost-model oblivious. We also show that as our cost-model estimation becomes crisp and stable the performance of learner improves significantly. Beyond the material presented in this paper, we continue studying the universality and robustness of the COREIL and rCOREIL approaches.

Specially for rCOREIL, it is an interesting problem to determine the optimal regularization parameter on the go or to adapt it with the dynamics of workload. Though now this process causes us only a one-time up-front cost, following the flavour of our approach we would like to perform it online. One possible method is to run COREIL for the first 500 queries and to calculate the costs for different set of regularization parameter values simultaneously for that period. Following that, we can choose the parameter value that causes minimum average estimation of the cost function.

We are now running further empirical performance evaluation tests with other datasets such as TPC-E, TPC-H and dedicated benchmarks for online index tuning [37]. For completeness from an engineering perspective, we are considering concurrent access, which was ignored in the algorithm and experiments presented in this paper for the sake of simplicity. We are also going to look at the favourable case of predictable workload such as periodic transactions. Furthermore, we are extending the solution to other aspects of database configuration, including partitioning and replication. For each of these aspects, we need to devise specific and non-trivial heuristics that help curb the combinatorial explosion of the configuration space as well as specific intelligent initialization techniques.

Finally, note that a critical assumption in our approach is that queries arrive sequentially and that nothing is known ahead of time about the workload. Both assumptions do not held in a number of realistic settings: queries can be submitted concurrently to the database, and a workload may often be predictable (such as when it consists of similar transactions, repeated on different data items). We leave for further work the adaptation of rCOREIL to such settings.

Acknowledgement. We thank Prof. Haibo Chen for valuable feedback on this work. This research is funded by the National Research Foundation Singapore under its Campus for Research Excellence and Technological Enterprise (CREATE) programme with the SP2 project of the Energy and Environmental Sustainability Solutions for Megacities – E2S2 programme.

References

1. Agrawal, S., Chaudhuri, S., Narasayya, V.R.: Automated selection of materialized views and indexes in sql databases. In: Proceedings of the 26th International Conference on Very Large Data Bases (VLDB 2000), pp. 496–505 (2000)
2. Agrawal, S., Narasayya, V., Yang, B.: Integrating vertical and horizontal partitioning into automated physical database design. In: Proceedings of the 2004 ACM SIGMOD International Conference on Management of Data (SIGMOD 2004), pp. 359–370 (2004)
3. Alagiannis, I., Idreos, S., Ailamaki, A.: H2o: a hands-free adaptive store. In: Proceedings of the 2014 ACM SIGMOD International Conference on Management of Data (SIGMOD 2014) (2014)
4. Audibert, J.Y., Munos, R., Szepesvári, C.: Exploration-exploitation tradeoff using variance estimates in multi-armed bandits. Theoret. Comput. Sci. **410**(19), 1876–1902 (2009)
5. Azefack, S., Aouiche, K., Darmont, J.: Dynamic index selection in data warehouses. CoRR abs/0809.1965 (2008). http://arXiv.org/abs/0809.1965

6. Basu, D., Lin, Q., Chen, W., Vo, H.T., Yuan, Z., Senellart, P., Bressan, S.: Cost-model oblivious database tuning with reinforcement learning. In: Chen, Q., Hameurlain, A., Toumani, F., Wagner, R., Decker, H. (eds.) DEXA 2015. LNCS, vol. 9262, pp. 253–268. Springer, Heidelberg (2015). doi:10.1007/978-3-319-22849-5_18

7. Benedikt, M., Bohannon, P., Bruns, G.: Data cleaning for decision support. In: Proceedings of the 1st International VLDB Workshop on Clean Databases (CleanDB 2006) (2006)

8. Borodin, A., El-Yaniv, R.: Online Computation and Competitive Analysis. Cambridge University Press, Cambridge (1998)

9. Bouchakri, R., Bellatreche, L., Hidouci, K.-W.: Static and incremental selection of multi-table indexes for very large join queries. In: Morzy, T., Valduriez, P., Bellatreche, L. (eds.) ADBIS 2015. LNCS, vol. 9282, pp. 43–56. Springer, Heidelberg (2012). doi:10.1007/978-3-642-33074-2_4

10. Bruno, N., Chaudhuri, S.: An online approach to physical design tuning. In: Proceedings of the 23th IEEE International Conference on Data Engineering (ICDE 2007), pp. 826–835 (2007)

11. Bruno, N., Chaudhuri, S.: Constrained physical design tuning. Proc. VLDB Endow. 1(1), 4–15 (2008)

12. Bruno, N., Chaudhuri, S.: Interactive physical design tuning. In: Proceedings of the 26th IEEE International Conference on Data Engineering (ICDE 2010), pp. 1161–1164 (2010)

13. Bruno, N., Nehme, R.V.: Configuration-parametric query optimization for physical design tuning. In: Proceedings of the 2008 ACM SIGMOD International Conference on Management of Data (SIGMOD 2008), pp. 941–952 (2008)

14. Chaudhuri, S., Narasayya, V.: Autoadmin: what-if index analysis utility. In: Proceedings of the 1998 ACM SIGMOD International Conference on Management of Data (SIGMOD 1998), pp. 367–378 (1998)

15. Difallah, D.E., Pavlo, A., Curino, C., Cudre-Mauroux, P.: Oltp-bench: an extensible testbed for benchmarking relational databases. Proc. VLDB Endow. 7(4), 277–288 (2013)

16. Gouriten, G., Maniu, S., Senellart, P.: Scalable, generic, and adaptive systems for focused crawling. In: Proceedings of the 25th ACM Conference on Hypertext and Social Media (HT 2014), pp. 35–45 (2014)

17. Hammer, M., Niamir, B.: A heuristic approach to attribute partitioning. In: Proceedings of the 1979 ACM SIGMOD International Conference on Management of Data (SIGMOD 1979), pp. 93–101 (1979)

18. Lagoudakis, M.G., Parr, R.: Least-squares policy iteration. J. Mach. Learn. Res. 4, 1107–1149 (2003)

19. Lai, T.L., Wei, C.Z.: Least squares estimates in stochastic regression models with applications to identification and control of dynamic systems. Ann. Stat. 154–166 (1982)

20. LeFevre, F., Sankaranarayanan, J., Hacigumus, H., Tatemura, J., Polyzotis, N., Carey, M.J.: Exploiting opportunistic physical design in large-scale data analytics. In: Proceedings of the 2014 ACM SIGMOD International Conference on Management of Data (SIGMOD 2014) (2014)

21. Li, L., Gruenwald, L.: Self-managing online partitioner for databases (smopd): a vertical database partitioning system with a fully automatic online approach. In: Proceedings of the 17th International Database Engineering and Applications Symposium (IDEAS 2013), pp. 168–173 (2013)

22. Lightstone, S., Bhattacharjee, B.: Automated design of multidimensional clustering tables for relational databases. In: Proceedings of the 30th International Conference on Very Large Data Bases (VLDB 2004), pp. 1170–1181 (2004)
23. Lohman, G.M.: Is query optimization a "solved" problem? (2014). http://wp.sigmod.org/?p=1075
24. Luhring, M., Sattler, K.U., Schmidt, K., Schallehn, E.: Autonomous management of soft indexes. In: Proceedings of the 2nd International Workshop on Self-Managing Data Bases (SMDB 2007), pp. 450–458 (2007)
25. Malik, T., Wang, X., Dash, D., Chaudhary, A., Ailamaki, A., Burns, R.: Adaptive physical design for curated archives. In: Ailamaki, A., Bowers, S. (eds.) SSDBM 2012. LNCS, vol. 7338, pp. 148–166. Springer, Heidelberg (2009). doi:10.1007/978-3-642-02279-1_11
26. Nielsen, F., Bhatia, R.: Matrix Information Geometry. Springer, Heidelberg (2013)
27. Papadomanolakis, S., Dash, D., Ailamaki, A.: Efficient use of the query optimizer for automated physical design. In: Proceedings of the 33rd International Conference on Very Large Data Bases (VLDB 2007), pp. 1093–1104 (2007)
28. Powell, W.B.: Approximate Dynamic Programming: Solving the Curses of Dimensionality. Wiley-Interscience, Hoboken (2007)
29. Puterman, M.L.: Markov Decision Processes Discrete Stochastic Dynamic Programming, vol. 414. Wiley, Hoboken (2009)
30. Raab, F.: TPC-C - the standard benchmark for online transaction processing (OLTP). In: Gray, J. (ed.) The Benchmark Handbook. Morgan Kaufmann, Burlington (1993)
31. Ramakrishnan, R., Gehrke, J., Gehrke, J.: Database Management Systems, vol. 3. McGraw-Hill, New York (2003)
32. Rao, J., Zhang, C., Megiddo, N., Lohman, G.: Automating physical database design in a parallel database. In: Proceedings of the 2002 ACM SIGMOD International Conference on Management of Data (SIGMOD 2002), pp. 558–569 (2002)
33. Rasin, A., Zdonik, S.: An automatic physical design tool for clustered column-stores. In: Proceedings of the 16th International Conference on Extending Database Technology (EDBT 2013), pp. 203–214 (2013)
34. Rieser, V., Robinson, D.T., Murray-Rust, D., Rounsevell, M.: A comparison of genetic algorithms and reinforcement learning for optimising sustainable forest management. GeoComputation (2011)
35. Rockafellar, R.T.: Convex Analysis. Princeton University Press, Princeton (2015)
36. Rösch, P., Dannecker, L., Färber, F., Hackenbroich, G.: A storage advisor for hybrid-store databases. Proc. VLDB Endow. 5(12), 1748–1758 (2012)
37. Schnaitter, K., Polyzotis, N.: A benchmark for online index selection. In: 2009 IEEE 25th International Conference on Data Engineering, pp. 1701–1708, March 2009
38. Schnaitter, K., Abiteboul, S., Milo, T., Polyzotis, N.: On-line index selection for shifting workloads. In: Proceedings of the 2nd International Workshop on Self-Managing Data Bases (SMDB 2007), pp. 459–468 (2007)
39. Schnaitter, K., Polyzotis, N.: Semi-automatic index tuning: keeping dbas in the loop. Proc. VLDB Endow. 5(5), 478–489 (2012)
40. Stillger, M., Lohman, G.M., Markl, V., Kandil, M.: LEO - DB2's LEarning Optimizer. In: VLDB (2001)
41. Sutton, R.S., Barto, A.G.: Reinforcement Learning. MIT Press, Cambridge (1998)

42. Warmuth, M.K., Jagota, A.K.: Continuous and discrete-time nonlinear gradient descent: relative loss bounds and convergence. In: Electronic proceedings of the 5th International Symposium on Artificial Intelligence and Mathematics. Citeseer (1997)
43. White, D.J.: Markov Decision Processes. Wiley, New York (1993)
44. Young, P.: Recursive least squares estimation. In: Recursive Estimation and Time-Series Analysis, pp. 29–46. Springer, Berlin, Heidelberg (2011)
45. Zilio, D.C., Zuzarte, C., Lightstone, S., Ma, W., Lohman, G.M., Cochrane, R., Pirahesh, H., Colby, L.S., Gryz, J., Alton, E., Liang, D., Valentin, G.: Recommending materialized views and indexes with IBM DB2 design advisor. In: Proceedings of the 1st International Conference on Autonomic Computing (ICAC 2004), pp. 180–188 (2004)

Workload-Aware Self-tuning Histograms
for the Semantic Web

Katerina Zamani[1], Angelos Charalambidis[1]([✉]), Stasinos Konstantopoulos[1],
Nickolas Zoulis[1,2], and Effrosyni Mavroudi[3]

[1] Institute of Informatics and Telecommunications,
NCSR 'Demokritos', Athens, Greece
{kzam,acharal,konstant}@iit.demokritos.gr
[2] Computer Science Department, Athens University of Economics
and Business, Athens, Greece
[3] School of Electrical and Computer Engineering, National Technical
University of Athens, Athens, Greece

Abstract. Query processing systems typically rely on histograms, data structures that approximate data distribution, in order to optimize query execution. Histograms can be constructed by scanning the database tables and aggregating the values of the attributes in the table, or, more efficiently, progressively refined by analysing query results. Most of the relevant literature focuses on histograms of numerical data, exploiting the natural concept of a numerical range as an estimator of the volume of data that falls within the range. This, however, leaves Semantic Web data outside the scope of the histograms literature, as its most prominent datatype, the URI, does not offer itself to defining such ranges. This article first establishes a framework that formalises histograms over arbitrary data types and provides a formalism for specifying value ranges for different datatypes. This makes explicit the properties that ranges are required to have, so that histogram refinement algorithms are applicable. We demonstrate that our framework subsumes histograms over numerical data as a special case by using to formulate the state-of-the-art in numerical histograms. We then proceed to use the Jaro-Winkler metric to define URI ranges by exploiting the hierarchical nature of URI strings. This greatly extends the state of the art, where strings are treated as categorical data that can only be described by enumeration. We then present the open-source STRHist system that implements these ideas. We finally present empirical evaluation results using STRHist over a real dataset and query workload extracted from AGRIS, the most popular and widely used bibliographic database on agricultural research and technology.

1 Introduction

Query optimizers in query processing systems typically rely on *histograms*, data structures that approximate data distribution, in order to be able to apply their cost model. Histograms can be constructed by scanning the database tables and aggregating the values of the attributes in the table; and similarly maintained in the face of database updates.

© Springer-Verlag Berlin Heidelberg 2016
A. Hameurlain et al. (Eds.): TLDKS XXVIII, LNCS 9940, pp. 133–156, 2016.
DOI: 10.1007/978-3-662-53455-7_6

This histogram lifecycle, however, cannot be efficiently applied to large-scale and frequently updated databases, such as, for example, stores of sensor data. An alternative approach is taken by *adaptive* query processing systems that update their histograms by observing and analysing the results of the queries that constitute the client-requested workload, as opposed to maintenance workload only for updating the histograms. The relevant databases literature focuses on numerical attributes, exploiting the concept of an *interval* as a description of a set of numerical values that is *succinct* and that has a *length* that can be used to estimate the cardinality of many different intervals that have roughly the same density.

In the work described here, we investigate how to extend adaptive query processing so that it can be applied to the domain of *strings*, typically treated as purely categorical symbols that can only be described by enumeration. This, however, disregards the fact that there are several classes of strings that have an internal structure and that can be handled in a more sophisticated manner. Specifically, we use string *prefixes* to expresses 'intervals', i.e., sub-spaces of the overall string space that are interesting from the point of view of providing query optimization statistics. Although weaker than regular expressions, prefixes can be very efficiently applied and can capture interesting ranges in hierarchically-structured string domains, such as that of URIs. We also experiment with describing a string range as a volume of strings similar to a central string, quantifying similarity in a way that favours similar prefixes.

This attention on URIs is motivated by their prominent position in the increasingly popular *Semantic Web* and *Linked Data* infrastructures for publishing data. In fact, these paradigms motivate adaptive query processing for a further reason besides the scale of the data: *distributed querying* engines often concentrate loose federations of publicly-readable remote data sources over which the distributed querying engine cannot effect that histograms are maintained and published. Furthermore, the URIs of large-scale datasets are not hand-crafted names but are automatically generated following naming conventions, usually hierarchical. These observations both motivate extending adaptive query processing to Semantic Web data stores and also present an opportunity for our string prefix extension.

In the remainder of this article, we first review self-tuning histograms (Sect. 2) where we identify STHoles as our starting point, a very successful algorithm for multi-dimensional histograms of numerical data. We proceed to formalize the key concepts in STHoles in a way that subsumes STHoles as its specialization for numerical intervals (Sect. 3) and to provide two alternatives for an extension that covers URI strings (Sect. 4). We then proceed to present experimental results using our prototype implementations (Sect. 5) and conclude (Sect. 6).

2   Background

In their simplest form, histograms describe an attribute a. The range of possible values of a is divided into non-overlapping *ranges*. A histogram is a set of *buckets* where each bucket is associated with a range and holds the number of tuples where the value for a is within the bucket's range. *Self-tuning* histograms are progressively refined from *query feedback* after each selection on a, using the actual result count to update the statistics in the bucket of a. In order to manage memory usage, some error is tolerated and buckets with similar statistics are merged into a single bucket with a wider range. In *workload-aware* self-tuning histograms, frequently used buckets (in a given workload) are split into narrower and more accurate buckets, while less frequently used buckets are more likely to be merged with more dissimilar buckets and produce a larger error when used.

Workload-aware self-tuning histograms have been successfully used in relational databases as a way to avoid the costly creation of static histograms of massive datasets. These techniques are memory efficient as they are focused towards the current workload, providing more accurate statistics for data regions that are being queried more frequently. Furthermore, they efficiently adapt to changes in the data distribution or the focus of the workload as they exploit query feedback collected from the production workload and do not impose any maintenance workload.

2.1 Histograms of Numerical Attributes

In one of their earliest instances [1], such one-dimensional histograms of numerical attributes were used to hold *statistics on intermediate tables (SIT)*, where each SIT corresponds to an intermediate node of the query plan. Adjacent buckets shared their ranges' edges and the ranges of all the buckets together covered the entire range of values of the attribute. In order to estimate the cardinality of arbitrary select-project-join (SPJ) queries, statistics are estimated for the individual patterns and then propagated through the query plan. Consider, for instance an SPJ query of the form:

$$(R.x = S.y) \text{ AND } (S.a < 10)$$

The histograms of tables R and S are used to estimate the selectivity of $R \bowtie S$ ignoring $S.a < 10$ and then the histogram of $S.a$ is used to estimate the selectivity of $S.a < 10$ over the result of $R \bowtie S$.

To avoid the propagation of errors through a sequence of operators, SITs cat also match intermediate sub-expressions of the query. That is to say, we would use statistics that are built on the result of the query expression $R \bowtie S$ specifically on ranges of $S.a$ values, rather than estimates derived from the isolated statistics of $R.x$, $S.y$, and $S.a$. A workload-driven technique was used to identify the SITs that maximized the benefit to the query optimizer.

These ideas are relevant to the Learning Optimizer (LEO) framework [2] used in DB2. LEO monitors query execution and accordingly adjusts the cardinality estimates and statistics used by the query optimizer. By comparing estimated and actual cardinalities, LEO gives positive or negative feedback to the statistics and the cardinality model used. Correlations can be also detected when estimates for individual predicates are known to be accurate but some combination of them is not. LEO does not modify statistics, but saves separately adjustment factors such that the product of the adjustment factor and the estimated selectivity derived from the DB2 statistics yields the correct selectivity. Stillger et al. [2] demonstrated that LEO improves cardinality estimates by orders of magnitude, changing plans to improve performance by orders of magnitude, while adding less than 5 % overhead to execution time when collecting query feedback.

STGrid [3] extends these ideas to multidimensional self-tuning histograms that use query workloads to refine a grid-based histogram structure. These self-tuning histograms are a low-cost alternative to traditional histograms with comparable accuracy. However, since the splitting (or merging) of each bucket entails the splitting (or merging) of several other buckets that could be far away from and unrelated to the original one, overall accuracy is degraded in order to satisfy the grid-partitioning constraint.

To alleviate the poor bucket layout problem of STGrid, STHoles [4] allows buckets to overlap. This more flexible data structure allows STHoles to exploit feedback in a truly multi-dimensional way and is adopted by many subsequent algorithms [5,6], including the one presented here. STHoles allows for inclusion relationships between buckets, resulting in a tree-structured histogram where each node represents a bucket. Holes are sub-regions of a bucket with different tuple density and are buckets themselves. To refine an STHoles histogram, query results are used to count how many tuples fall inside each bucket of the current histogram. Each partial intersection of query results and a bucket can be used to refine the histogram by drilling new holes, whenever the query results diverge from the prediction made through the bucket's statistics.

In order to maintain a constant number of buckets, buckets with close tuple densities are merged to make space for new holes. A penalty function measures the difference in approximation accuracy between the old and the new histogram to choose which buckets to merge. Parent-child merges are useful to eliminate buckets that become too similar to their parents; sibling merges are useful to extrapolate frequency distributions to yet unseen regions in the data domain and also to consolidate buckets with similar density that cover nearby regions.

ISOMER [5] is a more recent feedback-based algorithm for building and maintaining multidimensional histograms. ISOMER uses the histogram structure of STHoles and the information-theoretic principle of maximum entropy to refine the histogram based on *query feedback records (QFR)*. QFRs are $\langle q, N(q) \rangle$ records that match queries against the size of the query result. Once ISOMER obtains a consistent set of QFRs, the algorithm computes the 'simplest' (in terms of entropy) histogram that is consistent with all QFRs added so far. The result is a maximization problem under a system of constraints, solved with iterative scaling. Furthermore, to meet a space budget ISOMER discards QFRs merges buckets in a way similar to STHoles.

Except for storing cardinalities, another useful statistic for selectivity estimation of queries with equality or LIKE selection predicates is the number of distinct values. Kaushik and Suciu [7] presented the first self-tuning histogram modelling cardinalities and distinct value counts, which was based on the same *entropy maximization (EM)* principle as ISOMER but with a different probability space. Due to the computational complexity of the resulting EM problem, they minimize instead the squared distance between the histogram's estimates and the query feedback viewed as vectors. However, their method can only construct one-dimensional histograms on numerical or categorical data.

Markl et al. [8] address the problem of combining complementary selectivity estimations from multiple sources (estimations which are computed using ISOMER histograms) to obtain a consistent selectivity estimation using the idea of maximum entropy. Similar to the approach in ISOMER, this work exploits all available information and avoids biasing the optimizer towards plans for which the least information is known [9].

Khachatryan et al. [10] note that like traditional index structures such as R-Trees, STHoles fails in high-dimensional data spaces and is sensitive to the order of tree construction. As far as the latter is concerned, they argue that if the first few queries define a top-level bucket structure that is bad, the subsequent tuning is unlikely to correct it. They propose an initializing with subspace buckets which are derived from a subspace clustering algorithm. They use the MineClus cell clustering algorithm, which outputs a set of clusters with an assigned importance. Each cluster consists of tuples and has dimensions $d_1, d_2, ..., d_k$. The corresponding bucket is the minimal rectangle containing these points-tuples and spans the entire length of every dimension not in $d_1, d_2, ..., d_k$. They showed that the new initialization improves estimation quality and, in some situations, reduces the number of buckets.

2.2 Histograms of Categorical Attributes

STHoles and, in general, workload-aware self-tuning histograms have been successfully used in relational databases as a low-overhead alternative to statically re-scanning database tables. The resulting histogram is focused towards the current workload, providing more accurate statistics for data regions that are being queried more frequently. Furthermore, they are able to adapt to changes in data distribution and thus are well-suited for datasets with frequently changing contents. They are, however, for the most part targeting numerical attributes, since they exploit the idea that a value range is an indication of the size of the range. Turning our attention to the Semantic Web, the *Resource Description Framework (RDF)* is the dominant standard for expressing information. RDF information is a graph where *properties* (labelled edges) link two resources to each other or one resource to a *literal* (a concrete value). The relevance of this discussion to self-tuning histograms is that RDF uses URIs as abstract symbols that denote resources. Given this prominent role of URIs in RDF data, extending self-tuning histograms to string attributes can have a significant impact in optimizing querying of RDF datasets.

There has been relatively limited amount of work around string selectivity estimation in the field of relational databases. Chaudhuri et al. [11] proposed to collect multiple candidate identifying substrings of a string using, for example, a Markov estimator and build a regression tree as a combination function of their estimated selectivities, in order to alleviate the selectivity underestimation problem of queries involving string predicates in previous methods, which used independence and Markov assumptions. In 2005, Lim et al. [12] introduced CXHist, which is a workload-aware histogram for selectivity estimation supporting a broad class of XML string-based queries. CXHist is the first histogram based on classification that uses feature distributions to summarize queries and quantize their selectivities into buckets and a naive-Bayes classifier to capture the mapping between queries and their selectivity.

Within the Semantic Web community itself, the SWOOGLE search engine collects metadata, such as classes, class instances and properties for web documents and relations between documents [13]. LODStats computes several schema-level statistical values for large-scale RDF datasets using an approach based on *statement streams* [14]. More closely related to our work is RDFStats [15], which is a generator for statistics of RDF sources like SPARQL endpoints. They generate different statistical items such as instances per class and histograms. Unlike our approach, they generate different static histograms (i.e. that must be rebuilt to reflect any changes in the RDF source) per class, property and XML data type. For range estimations on strings, RDFStats mentions three possibilities: (a) one bucket for each distinct string, resulting in large histograms; (b) reducing strings to prefixes; or (c) using a hash function to reduce the number of distinct strings, although no appropriate general-purpose hash function has been identified. However, as Harth et al. [16] have also noted in relation to Q-Trees for indexing RDF triples, hashing URIs is a purely syntactic mapping from URIs to numerical coordinates and fails to take into account the semantic similarity between resources; and no universally good function has been identified.

As URIs are the most prominent datatype in the Semantic Web, the absence of an extension that can naturally handle URI strings leaves Semantic Web data outside the scope of many developments in self-tuning histograms.

3   Self-Tuning String Histograms

In this section we establish a new histogram structure that extends the structure of the STHoles algorithm with the ability to cover strings. We also present the algorithms that construct and refine this new structure.

In our treatment, we first defer defining how string ranges are specified. Instead, we construct a framework of preliminary definitions where we specify the properties that must be satisfied by any compliant definition of string ranges. We then proceed to construct two alternative string ranges: the first one is based on prefixes and is a slight re-formulation of previous work [17] so that it complies with this framework. The second definition of string ranges is based on string distance.

3.1   Preliminaries

Let D be a *dimension*, any subset of D be a *range* in D, and $\mathcal{P}(D)$ the set of all possible ranges in D. A range can be defined either implicitly by constraints over the values of D or explicitly by enumeration. Note that $D \in \mathcal{P}(D)$, meaning that a range does not *need* to impose a restriction but can also include the whole dimension. Let H be a histogram of n dimensions $D_1, \ldots D_n$. Let $V(H)$ be the set of all possible n-dimensional vectors $(r_1, \ldots r_n)$ where $\forall i \in [1, n] : r_i \in \mathcal{P}(D_i)$.

A histogram is represented as an inclusion hierarchy of *buckets*; we shall use B_H to denote the set of buckets of a histogram H.

Definition 1. *Each bucket $b \in B_H$ is an entity of histogram H such that:*

- *b is associated with a box$(b) \in V_H$, the vector that specifies the set of tuples that the bucket describes.*
- *b is associated with a size(b) which indicates the number of tuples that match box(b)*
- *b is associated with n values dvc$(b, D_i), i = 1 \ldots n$ which indicate the number of distinct values appearing in dimension D_i of the tuples that match box(b).*

We define the density of a bucket b to be the quantity

$$\text{density}(b) = \frac{\text{size}(b)}{\prod\limits_{i : r_i \in \text{box}(b)} \text{dvc}(b, D_i)}$$

Definition 2. *Every histogram implicitly includes a bucket $b_\top$ such that* box$(b_\top) \equiv (D_1, \ldots D_n)$ *that is, the bucket that imposes no restrictions in any of the dimensions of H and includes all tuples. We call this the* top bucket $b_\top$.

The implication of Definition 2 is that the overall size of the dataset and the number of distinct values in each dimension should be known (or at least approximated) regardless of what query feedback has been received. In our implementation we assume the *root bucket* (the top-most bucket of the hierarchy) as an approximation of the top.

Let $\mathcal{Q}_H$ be the set of all possible queries over the tables covered by H. Regardless of how they are syntactically expressed, we perceive $\mathcal{Q}_H$ as the set of all possible restrictions over the dimensions of H; thus:

Definition 3. *Each query $q \in \mathcal{Q}_H$ is an entity of histogram H such that:*

- *q is associated with a box$(q) \in V_H$, the vector that specifies the restrictions expressed by the query*
- *q is associated with a size(q) which indicates the number of tuples that are returned by executing q.*

As pointed out earlier, in our preliminary constructions ranges are simply defined as any subset of D, without making any requirements on how these are specified; in fact they may even be specified by enumeration. However, in order to realize the memory efficiency of workload-aware histograms, ranges should be specified intensionally, so that their representation consumes a memory unit regardless of how many elements match the specification. We shall present below the definitions we propose for string ranges; at this point, it suffices to define a range as follows:

Definition 4. *We define a* range $r \in \mathcal{P}(D)$ *of dimension* D *of histogram* H *to be an entity of* H *with the following properties:*

- *There is a* membership function $member_r : D \to \{true, false\}$ *that can consistently decide for any* $t \in D$ *whether it is or is not inside* r.
- *There is an* intersection function $\text{\footnotesize ⋒} : \mathcal{P}(D) \times \mathcal{P}(D) \to \mathcal{P}(D)$ *that returns the range resulting from the intersection of two ranges.*

It should be noted that we do not make any claims on the intersection function, although it is advantageous if such a function is approximately (if not exactly) the same as a function that would output a range that has as extension the intersection of the ranges' extensions. However, it should be possible to operate in instantiations of the framework where the (strict) intersection cannot be computed or syntactically represented for all pairs of ranges. In such instantiations, our relaxed definition of $\text{\footnotesize ⋒}$ provides the flexibility to define operators that roughly (but not exactly) correspond to producing a representation for the intersection of the extensions of its operands.

We use range intersection to also define multi-dimensional box intersection as follows:

Definition 5. *Given two boxes* $v_1, v_2 \in V_H$ *from the* n-*dimensional histogram* H, *let* $v_1 = (r_{1,1}, \ldots r_{1,n})$ *and* $v_2 = (r_{2,1}, \ldots r_{2,n})$. *We define* box intersection:

$$v_1 \cap v_2 = (r_{1,1} \cap r_{2,1}, \ldots r_{1,n} \cap r_{2,n})$$

Definition 6. *Given two boxes* $v_1, v_2 \in V_H$ *from the* n-*dimensional histogram* H, *let* $v_1 = (r_{1,1}, \ldots r_{1,n})$ *and* $v_2 = (r_{2,1}, \ldots r_{2,n})$. *We say that* v_1 encloses v_2 *iff* $\forall i \in [1, n]$ *at least one of the following holds:*

1. $r_{2,i} \subseteq r_{1,i} \subset D_i$, *that is, none of the ranges is the complete dimension and* $r_{2,i}$ *is contained within* $r_{1,i}$
2. $r_{2,i} = D_i$ *and* $r_{1,i} \subset D_i$, *that is, if one of the ranges is the complete dimension then it is enclosed by the one that is not.*
3. $r_{2,i} = r_{1,i} = D_i$, *that is, both ranges are the complete dimension.*

It should be noted that we have defined an unrestricted dimension as *being enclosed by* (rather than enclosing) a restriction. The rationale behind this will be explained in conjunction with *bucket merging* (Sect. 3.4).

Definition 7. *Given two boxes* $v_1, v_2 \in V_H$ *from histogram* H, v_1 *tightly encloses* v_2 *iff* v_1 *encloses* v_2 *and there is no* $u \in V_H$ *such that* $v_1 \supsetneq u \supsetneq v_2$.

Definition 8. *Given a query* $q \in \mathbb{Q}_H$, *we associate with* q *the* best fit, *the set of buckets* $\text{bf}(q) \subseteq B_H$ *such that*

$$\forall b \in \text{bf}(q) : \text{box}(b) \text{ tightly encloses } \text{box}(q)$$

Lemma 1. *For every query there is always a non-empty best fit.*

Proof. There is always at least one bucket that *encloses* any box(q), the *top bucket* $b_\top$ (Definition 2). If there is no other bucket that *encloses* box(q), then $b_\top$ *tightly encloses* box(q) (Definition 7) and thus bf(q) = $\{b_\top\}$, which is non-empty. If there are other buckets that *enclose* box(q), then there is also at least one that *tightly encloses* box(q), so bf(q) is non-empty.

3.2   Cardinality Estimation

Being able to predict the size of querying results is important input for query execution optimizers, but the specifics of how this optimization is performed is outside the scope of this paper. We will here proceed to define metrics over the values associated with the buckets of H in order to predict size(q), $q \in \mathcal{Q}_H$, the number of results returned by q.

In the literature, numerical intervals are used to succinctly define ranges and to efficiently decide if a query is enclosed by a bucket or not. The numerical difference between the interval's starting and ending value is sometimes used to define *range length* and, in multi-dimensional buckets, *bucket volume*: an estimator of the number of tuples in a bucket. We, accordingly, define range length as follows:

Definition 9. *Given a histogram dimension D and a range $r \in \mathcal{P}(D)$ we define the function* length $: \mathcal{P}(D) \rightarrow \mathbb{R}$ *as follows:*

1. *Unrestricted ranges that span the whole dimension have length 0.*
2. *If r is an extensionally defined range of any type, then* length(r) = $|r|$, *the number of distinct values in the range.*
3. *If r is a numerical range defined by an interval $[x, y]$, then* length(r) = $y - x + 1$.

The addition of the unit term guarantees that the length cannot be zero even if $x = y$, i.e., even if the numerical range is a single point. This makes the third clause of the definition consistent with the second one, since for any number n we would expect the length of the singleton $\{n\}$ according to clause 2 to be the same as the length of the range $[n, n]$ that can also only include a single distinct value. It should also be noted that this is the only situation in which the length of a range can be 1. This property is important for Definition 10 below.

We will revisit this definition in Sect. 4 and complete it with the definition of length for URI ranges. Regardless of how length is defined for numerical, URI, or other types of ranges, we propose the following function as an estimator of the number of tuples that lie inside q, given a histogram:

Definition 10. *Given a histogram H and a query q, let* box(q) = $r_1, ... r_n$. *We define the function* est$_H : V_H \rightarrow \mathbb{R}$ *as follows:*

$$\text{est}_H\left(\text{box}(q)\right) = \sum_{b \in bf(q)} \frac{\text{size}(b)}{\prod_{i:length(r_i)=1} \text{dvc}(b, D_i)}$$

The intuition behind this definition is that we identify a best-fitting bucket (cf. Definition 8) and assume that tuples are uniformly distributed among the distinct values in each dimension. Since the query might have bindings for some of its dimensions, we use this assumption to apply simple division to estimate the fraction of the bucket's tuples that will be selected by the query dimensions that are unbinded variables. Naturally, this also assumes that the length of the range of the query's box can only be 1 for binded dimensions and is greater than 1 otherwise. This property is guaranteed by the definition of categorical and numerical length (Definition 9), and should also be observed by any extensions for other types.

3.3 Histogram Construction and Refinement

The construction of the histogram follows the same high level steps as the STHoles algorithm. In particular, we start with an empty histogram. For each query q in the workload, we identify *candidate buckets* b_i that intersect with q. For each candidate bucket b_i we compute $b_i \sqcap q$ and these intersections constitute *candidate holes* c_i. We then shrink each candidate hole to the largest sub-region that does not intersect with the box of any other bucket, we count the exact number of tuples from the result stream that lie inside the shrunk hole and the distinct values count. Then, we determine whether the current density of the candidate bucket is close to the actual density of the candidate hole. If not, we 'drill' the candidate hole as a new histogram bucket and we move all children of b_i that are enclosed by c_i to the new bucket (Algorithms 1 and 2).

A point of divergence from STHoles is when shrinking candidate holes. Let X be the set of all buckets that partially intersect with a candidate hole c_i. STHoles selects at each step the pair $\langle x, j \rangle$ that comprises bucket $x \in X$ and dimension j such that shrinking c_i along j by excluding x has as a result the smallest reduction of c_i. Instead of checking for the optimal $\langle x, j \rangle$ our method selects the first pair where shrinking c_i along j by excluding x results in the smallest *relative* reduction of c_i's *length* in that dimension, the intuition being that often excluding x will give similar relative reduction along all dimensions.

Algorithm 1. Refinement of a histogram H given a set of queries W.

---

procedure REFINE(H,W)
    for all queries $q \in W$ **do**
        if q is not contained in H **then**
            expand H's root bucket so that it contains q
        for all buckets b_i such that $q \sqcap b_i \neq \emptyset$ **do**
            $(c_i, T_{c_i}, d_{c_i}) \leftarrow$ SHRINKBUCKET(b_i, q)
            if estimation is not accurate **then**
                DRILLHOLE($b_i, c_i, T_{c_i}, d_{c_i}$)
    while H has too many buckets **do**
        Let b_1, b_2 in H with the lowest penalty$_H(b_1, b_2)$
        MERGE(b_1, b_2)

Algorithm 2. Drilling a hole in bucket b, given a candidate hole c and the counted cardinality T_c and distinct values $D_c(i)$ for each dimension D_i.

---

   procedure DRILLHOLE(b, c, T_c, $d_c(\cdot)$)
      if box(b) = box(c) **then**
         size(b) ← T_c
         dvc(b, D_i) ← $D_c(i)$ $\forall i \in attributes$
      else
         Add a new child b_n of b to the histogram
         box(b_n) ← c
         size(b_n) ← T_c
         dvc(b_n, D_i) ← $d_c(i)$ $\forall i \in attributes$
         Migrate all children of b that are enclosed by c
         so they become children of b_n

Algorithm 3. Shrink a bucket that is enclosed by the intersection of b and q and does not partially intersect any other bucket.

---

   function SHRINKBUCKET(b, q)
      $c \leftarrow$ box(q) $\cap$ box(b)
      $\mathcal{P} \leftarrow \{b_i \in children(b) \mid c \cap \text{box}(b_i) \neq \emptyset \wedge \text{box}(b_i) \not\subseteq c\}$
      while $\mathcal{P} \neq \emptyset$ **do**
         Get first bucket $b_i \in \mathcal{P}$ and dimension j
         such that shrinking c along j by excluding b_i results
         in the smallest reduction of c.
         Shrink c along j
         $\mathcal{P} \leftarrow \{b_i \in children(b) \mid c \cap \text{box}(b_i) \neq \emptyset \wedge \text{box}(b_i) \not\subseteq c\}$
      Count from the result the number of tuples in c, T_c
      for all attributes i **do**
         Count from the result the number of
         distinct values of the ith attribute in c, $d_c(i)$.
   return $(c, T_c, d_c(\cdot))$

We then shrink c_i, we update participants and repeat the procedure until there are no participants left (Algorithm 3). This may result in a suboptimal shrink, but we avoid examining all possible combinations at each step. Furthermore, in STHoles the number of tuples in this shrunk subregion is estimated assuming uniformity; instead, we measure exactly the number of tuples and distinct values per dimension.

3.4 Bucket Merging

In order to limit the number of buckets and memory usage, buckets are *merged* to make space for drilling new holes. Following STHoles, our method looks for *parent-child* or *sibling* buckets that can be merged with minimal impact on the cardinality estimations. We diverge from STHoles when computing the box, size, and dvc associated with the merged bucket as well as in the *penalty* measure that guides the merging process towards merges that have the smallest impact on estimation accuracy.

Let b_1, b_2 be two buckets in the n-dimensional histogram H and let H' be the histogram after the merge and b_m the bucket in H' that replaces b_1 and b_2. In the *parent-child* case, the parent bucket, let that be b_1, *tightly encloses* the child bucket. In this case, we merge b_2 into b_1, so that $\text{box}(b_m) \equiv \text{box}(b_1)$. Any children that b_2 had become children of b_m.

In *sibling-sibling* merges, let b_p be the common parent bucket that *tightly encloses* both siblings b_1 and b_2. The merged bucket b_m is a child of b_p and the parent of all children of b_1 and b_2. The box of b_m must be such that it encloses the boxes of b_1 and b_2, without partially overlapping with any further siblings. Different implementations might achieve this either by defining checks that block sibling merges or by defining the box of b_m in such a way that it also encloses any further siblings that partially overlap with the extended box that encloses b_1 and b_2.

The size of b_m is estimated by adding the sizes of b_1 and b_2; the distinct values count of b_m is estimated by the maximum distinct values count among the merged buckets:

1. $\text{box}(b_p)$ *tightly encloses* $\text{box}(b_m)$
2. $\text{box}(b_m)$ *tightly encloses* both buckets b_1, b_2
3. $\text{box}(b_m)$ *tightly encloses* the boxes of all children of b_p that partially intersect either of b_1, b_2. That is, $\text{box}(b_m)$ encloses $\text{box}(b_c)$ for all b_c such that:
   (a) b_p *tightly encloses* b_c; and
   (b) $\text{box}(b_1)$ *partially overlaps* $\text{box}(b_c)$ or $\text{box}(b_2)$ *partially overlaps* $\text{box}(b_c)$
4. $\text{size}(b_m) = \sum\limits_{k=1,2,c_1,\ldots} \text{size}(b_k)$
5. $\text{dvc}(b_m) = \max\limits_{k=1,2,c_1,\ldots} \text{dvc}(b_k)$

It should be noted that the procedure that constructs the merged bucket b_m is deterministic and thus b_m can be uniquely determined by b_1 and b_2. In Point 3 above, it should be stressed that the partially intersecting buckets b_c are *not* merged into b_m, but that the latter is expanded so that it can assume b_c as its children. This is because in some algorithms (including STHoles), $\text{box}(b_m)$ can become larger than $\text{box}(b_1) \cup \text{box}(b_2)$ in order to have a succinct description with a single interval in each dimension. As a result, it might cut across other buckets; $\text{box}(b_m)$ should then be extended so as to subsume those as children. In order to avoid, however, dropping informative restrictions, STHoles only extends $\text{box}(b_m)$ along dimensions where the boxes of b_c do have a restriction. In order to capture this, we have defined the *encloses* relation (Definition 6) in a way that makes unrestricted dimensions *enclosed by* (rather than enclosing) restrictions.

In order to decide which is the optimal merge at any stage of histogram refinement, we need to balance between merges of buckets with similar statistics (minimizing the error introduced by discarding the statistics held in the merged buckets) and buckets with similar boxes (minimizing the error introduced by generalizing boxes beyond what was warranted by hole drilling, i.e., query feedback). To achieve the latter, we first define a *distance function* that evaluates

Definition 11. *Given a histogram H and any two of its boxes v_1 and v_2, we define the* distance *between v_1 and v_2 as any function* $\text{distance}_H : V_H \times V_H \to \mathbb{R}$ *that has the following properties:*

- $\text{distance}(v_1, v_1) = 0$
- *If v_1 encloses v_2 then* $\text{distance}(v_1, v_2) = 0$.

We can now define the *penalty function* that evaluates a possible merge:

Definition 12. *Given a histogram H and any three of its buckets b_1, b_2 and b_m, we define the* penalty function $\text{penalty}_H : B_H \times B_H \to \mathbb{R}$ *of merging b_1 and b_2 into b_m as follows:*

$$
\begin{aligned}
\text{penalty}_H\,&(b_1, b_2) \\
= \frac{1}{2} &\left(\frac{|\text{density}(b_1) - \text{density}(b_m)|}{\text{density}(b_1) + \text{density}(b_m)} + \frac{|\text{density}(b_2) - \text{density}(b_m)|}{\text{density}(b_2) + \text{density}(b_m)} \right) \\
+ \sum_i &\left(\frac{|\text{dvc}(b_1, i) - \text{dvc}(b_m, i)|}{\text{dvc}(b_1, i) + \text{dvc}(b_m, i)} + \frac{|\text{dvc}(b_2, i) - \text{dvc}(b_m, i)|}{\text{dvc}(b_2, i) + dvc(b_m, i)} \right) \\
+ \text{distance}\,&(\text{box}\,(b_1)\,, \text{box}\,(b_2))
\end{aligned}
$$

The first two terms of this function represent the error in the statistics introduced by the merge while the third term increases the penalty for bucket pairs that are more distant as defined in Definition 11. Therefore, a sibling-sibling merge must have a small enough statistics-based penalty to be preferred over a parent-child merge, so that it can counter the fact that parent-child merges always have 0 distance-based penalty (since a child is always enclosed by its parent).

This penalty function allows us to rank the candidate bucket pairs and select the one with the minimum penalty. It should be noted though that not every bucket pair can be candidate for merging. The following merging constraints apply:

- The new $box(b_m)$ should not intersect with any other box, otherwise we would result in an inconsistent histogram
- The new $box(b_m)$ should not cover more than the half volume of its parent. This constraint is significant in order to control over-generalization in the early stages of an histogram when distant siblings might not be blocked from merging by the previous clause
- If the new $box(b_m)$ encloses the boxes of other buckets, b_m assumes these buckets as as its children.

The specifics of how to calculate the box of the merged bucket are left to be defined for each dimension type.

3.5   Extending for Further Types

We have deliberately avoided binding the discussion so far to specific data types, in order to define a general framework for histograms. The only exception is that the *length* of numerical ranges is already defined (Definition 9), in order to ensure backwards compatibility with numerical ranges in STHoles.

In order to specify the histograms of a new data type, which we shall here call *newtype*, one needs to provide the following:

1 A function *newtype member* that satisfies the definition of the generic *member* function (Definition 4).
2. A function *newtype intersection* that satisfies the definition of the generic *intersection* function (Definition 4).
3. A function *newtype length* that satisfies the definition of the generic *length* function (Definition 9).
4. A function *newtype distance* that satisfies the definition of the generic *distance* function (Definition 11).
5. A procedure for calculating the box of the resulting bucket in sibling merging. This procedure must satisfy the merging constraints in Sect. 3.4.

In the following section we will proceed to present two alternative specifications for URI histograms within this framework.

4   URI Ranges

As a first approach to expressing ranges of URIs, we have looked at prefixes. Prefixes can naturally express ranges of semantically related resources given the natural tendency to group together relevant items in hierarchical structures such as pathnames and URIs. We have also experimented with exploiting a geometrical analogy where we express a range as the volume around a central URI; again, we have defined distance in a way that prefixes weigh more, in order to preserve the bias towards hierarchical structures but offering more flexibility by comparison to exact prefix matching.

4.1   Prefix Ranges

In this approach we assume string prefixes as the description language for implicitly defining string ranges.

Definition 13. *Let H be a histogram and D be a string dimension of H. We define a* prefix range r *of D to be a set of strings, denoted as* $\mathrm{Pref}(r)$. *The strings in* $\mathrm{Pref}(r)$ *are to be interpreted as the prefixes of the elements of D that are in r. For any string $s \in D$ we define* prefix membership *as follows:*

$$\mathrm{member}_r(s) = \begin{cases} true, & \exists p \in \mathrm{Pref}(r) : s \text{ starts with } p \\ false, & otherwise \end{cases}$$

In order to satisfy the requirements set in Sect. 3.5, we need to define the functions *prefix intersection*, *prefix length*, and *prefix distance* over prefix ranges, as well as the procedure for sibling merging.

Definition 14. *Let H be a histogram, D a string dimension of H, and $r_1, r_2 \in \mathcal{P}(D)$ two prefix ranges over D. The range intersection $r_1 \cap\!\!\!\!\!\!\!\! \text{m} \,\, r_2$ is defined as:*

1. *If r_1, r_2 are string ranges defined by sets of prefixes, then $r_1 \,\text{⋒}\, r_2 = \{p | (p_1, p_2) \in \text{Pref}(r_1) \times \text{Pref}(r_2) \wedge (p = p_1 = p_2 \vee \text{ one of } p_1, p_2 \text{ is a prefix of the other and } p \text{ is the longest (more specific) of the two})\}$*
2. *If one of the ranges is a string range defined by sets of prefixes (say r_1 without loss of generality) and the other is an explicit set of strings (say r_2), then $r_1 \,\text{⋒}\, r_2 = \{v | v \in r_2 \wedge \exists p \in r_1 : p \text{ is a prefix of } v\}$*
3. *In any other case, $r_1 \,\text{⋒}\, r_2 = r_1 \cap r_2$.*

Definition 15. *Given a histogram dimension D and a range $r \in \mathcal{P}(D)$ we define the function $\text{length} : \mathcal{P}(D) \rightarrow \mathbb{R}$ as follows:*

1. *Unrestricted ranges that span the whole dimension have length 0.*
2. *If r is an extensionally defined range of any type, then $\text{length}(r) = |r|$, the number of distinct values in the range.*
3. *If r is a numerical range defined by an interval $[x, y]$, then $\text{length}(r) = y - x + 1$.*
4. *If r is a string range defined by a set of prefixes $\text{Pref}(r)$, then $\text{length}(r) = 1 + |\text{Pref}(r)|$*

It should be noted that no prefix range can ever be guaranteed to be equivalent to an extensional singleton range, since any valid URI prefix can be extended into a longer valid URI subsumed by the prefix. Therefore, all and only extensional singleton ranges can have a length of 1, which satisfies Requirement 3.

Definition 16. *Let r_1 and r_2 be prefix ranges. We define the prefix distance between r_1 and r_2 to be a constant 0 for any r_1, r_2.*

That is to say, in this setup there is no bias in sibling merges towards more similar prefixes and candidate merges are evaluated only on the basis of the similarity of the statistics in the buckets.

Box of Merged Siblings. Suppose that sibling buckets b_1 and b_2 are to be merged. The box of the merged bucket b_m is calculated as the union of the prefixes in each:

$$\text{Pref}(\text{box}(b_m)) = \text{Pref}(\text{box}(b_1)) \cup \text{Pref}(\text{box}(b_2))$$

4.2 Similarity Ranges

In this approach we use the Jaro-Winkler similarity metric [18] to define the distance between two strings. This metric is suitable for URI comparison since it provides preference to the strings that match exactly at the beginning. Based on this, we define URI ranges as spherical volumes around a characteristic central URI, so that a range is specified by a URI (the center) and the radius around it that is within the range.

Definition 17. *Let H be a histogram and D be a string dimension of H. Let* $JW : D \times D \to [0,1]$ *be the Jaro-Winkler metric that assigns a similarity to an unordered pair of strings from D. We define similarity range* r_d *as a tuple* $r_d = \langle c, R \rangle$ *where c is a string called the* center *of r denoted as* center(r) *and* $R \in \mathbb{R}$ *is called the* radius *of r and denoted as* radius(r). *For any string* $s \in D$ *we define* similarity membership *as follows:*

$$\mathrm{member}_r(s) = \begin{cases} true, & if\ 1 - \mathrm{JW}(s, center(r)) \le radius(r) \\ false, & otherwise \end{cases}$$

In order to satisfy the requirements set in Sect. 3.5, we need to define the functions *similarity intersection, similarity length,* and *similarity distance* over similarity ranges, as well as the procedure for sibling merging.

Definition 18. *Given two similarity ranges of the same dimension* $r_1, r_2 \in \mathcal{P}(D)$ *their* similarity intersection *is defined as* $r_1 \cap r_2 = \langle c', R' \rangle$ *where:*

$$c' = center(r_i)\ where\ i = \mathrm{argmax}_{i=1,2}\ radius(r_i)$$
$$R' = \max\{0, radius(r_1) + radius(r_2) - \mathrm{distance}_H(r_1, r_2)\}$$

Definition 19. *Given a histogram dimension D and a range* $r \in \mathcal{P}(D)$ *we define the* similarity length *function* length $: \mathcal{P}(D) \to \mathbb{R}$ *as follows:*

1. *Unrestricted ranges that span the whole dimension have length 0.*
2. *If r is an extensionally defined range of any type then* length(r) = |r|, *the number of distinct values in the range.*
3. *If r is a numerical range defined by an interval* [x, y], *then* length(r) = y − x + 1.
4. *If r is a similarity range then* length(r) = 1 + radius(r)

It should be noted that range $\langle u, 0 \rangle$ has u as its single member and is equivalent to the extensional singleton range u. The similarity range length is 1 in both cases, which satisfies Requirement 3.

Definition 20. *Let* r_1 *and* r_2 *be similarity ranges. We define the* similarity distance *between* r_1 *and* r_2 *using the Jaro-Winkler similarity of their centers:*

$$\mathrm{distance}_H(r_1, r_2) = 1 - \mathrm{JW}(center(r_1), center(r_2))$$

Box of Merged Siblings. Suppose that sibling buckets b_1 and b_2 are to be merged. The box of the merged bucket b_m is calculated for each dimension i that is URI dimension, where r_1 is the range of b_1 in dimension i, r_2 is the range of b_2 in dimension i, and r_m is the range of b_m in dimension i. We assume that for every range r_i we can assign consistently an id and without loss of generality let r_1 be the range with the smallest id.

1. If $radius(r_1) = 0$ and $radius(r_2) = 0$, then:

$$center(r_m) = center(r_1)$$
$$radius(r_m) = \mathrm{distance}_H(r_1, r_2)$$

2. If $radius(r_1) \neq 0 \wedge radius(r_2) \neq 0$, then:

$$center(r_m) = \begin{cases} center(r_1), & \text{if } radius(r_1) \geq radius(r_2) \\ center(r_2), & \text{otherwise} \end{cases}$$

$$radius(r_m) = \begin{cases} \text{distance}_H(r_1, r_2) + radius(r_2), & \text{if } radius(r_1) \geq radius(r_2) \\ \text{distance}_H(r_1, r_2) + radius(r_1), & \text{otherwise} \end{cases}$$

3. otherwise,

$$center(r_m) = \begin{cases} center(r_1), & \text{if } radius(r_1) \neq 0 \\ center(r_2), & \text{otherwise} \end{cases}$$

$$radius(r_m) = \text{distance}_H(r_1, r_2)$$

That is, the center of the merged range is that of the range with the greater radius, and the radius of the merged range is large enough so that the merged range also encloses the range with the smaller radius. The intuition behind this definition is that by assuming the larger of the two ranges as the basis for the merged range, a smaller expansion will be needed in order to enclose the other range, reducing the risk of over-generalizing.

4.3   Discussion

We have defined a multi-dimensional histogram over numerical, string, and categorical data. The core added value of this work is that we introduce the notion of *descriptions* in string dimensions, akin to intervals for numerical dimensions. This has considerable advantages for RDF stores and, more generally, in the Semantic Web and Linked Open Data domain, where URIs have a prominent role and offer the opportunity to exploit the hierarchical structure of their string representation.

Initially, we propose *prefixes* as the formalism for expressing string ranges, motivated by its applicability to URI structure. We then relax this formalism, using *similarity ranges* to describe string ranges based on string distances. This is no loss of generality, since it is straightforward to use more expressive pattern formalisms (such as regular expressions) without altering the core method but at a considerable computational cost. The only requirement is that *membership*, *intersection* and some notion of *length* can be defined. Length, in particular, can be used in the way STHoles uses it as an indication of a bucket's size relative to the size of its parent bucket. If a metric of *distance* or *dissimilarity* can be defined, this is also exploited to introduce bias towards merging similar ranges, but this is not required.

What allows us to relax the definition of length by comparison to STHoles, is that for range queries we return the statistics of the bucket that more tightly encloses the query, instead of returning an estimation based on the ratio of the volume occupied by the query to the volume of the overall bucket. In other words, we use *length* more as a metric of the size of description, rather than a metric of the bucket size (the number of tuples that fit this description). To compensate, we exactly measure in query results (rather than estimate) bucket size when shrinking buckets, compensating for the extra computational time by avoiding examining all combinations of buckets × dimensions (cf. Sect. 3.3). For point queries (with unit length), we also take into account statistics about distinct value counts in a bucket, increasing the accuracy of the estimation.

A limitation of our algorithm is that when we merge two sibling buckets we assign to the resulting bucket the sum of the sizes of the merged buckets and of the children of the resulting bucket, which is an overestimation of the real size. Furthermore, we also assign as distinct value count the maximum of the distinct value counts of these buckets, which is an underestimation of the real distinct value count. These estimations will persist until subsequent workload queries effect an update of merged bucket's statistics and will be used in cardinality estimations. We try to compensate for these possibly inaccurate estimations by carefully selecting buckets for sibling-sibling merging and defining a sibling-sibling merge penalty which favours the merging of buckets which not only have similar statistics, i.e. densities and distinct value counts, but their central strings are also similar. Besides empirically testing and tuning these estimators, we are also planning to extend the theoretical framework so that estimated values are represented as ranges or distributions, and subsequent calculations take into account the whole range or the distribution parameters rather than a single value.

In general, and despite these limitations, our framework is an accurate theoretical account of STHoles, a state-of-the-art algorithm for self-tuning multi-dimensional numerical histograms, and an extension to heterogeneous numerical/string histograms that is backwards-compatible with STHoles.

5   Experiments

To empirically validate our approach, the algorithm presented above has been implemented in Java as the *STRHist* module of the *Semagrow Stack* [19], an optimized distributed querying system for the Semantic Web.[1] The execution flow of the Semagrow Stack starts with client queries, analysed to build an optimal *query plan*. The optimizer relies on cardinality statistics (produced by STRHist) in order to provide an execution plan for the Semagrow *Query Execution Engine*. This engine, besides joining results and serving them to the client application, also forwards to STRHist measurements collected during query execution. STRHist analyses these query feedback logs in batches to maintain the histogram that is used by the optimizer. The histogram is persisted in RDF stores using the Sevod vocabulary [20], which expresses the in-memory tree of *bucket* objects that is the internal representation of STRHist.

---

[1] *STRHist* is available at https://github.com/semagrow/strhist. For more details on Semagrow, please see http://semagrow.github.io.

5.1   Experimental Setup

We applied STRHist to the AGRIS bibliographic database on agricultural research and technology maintained by the *Food and Agriculture Organization of the UN*. AGRIS comprises approximately 23 million RDF triples describing 4 million distinct publications with standard bibliographic attributes.[2] AGRIS consolidates data from more than 150 institutions from 65 countries. Bibliography items are denoted by URIs that are constructed following a convention that includes the location of the contributing institution and the date of incorporation into AGRIS. As scientific output increases through the years and since there is considerable variation in the scientific output of different countries, there are interesting generalizations to be captured by patterns over publication URIs.

We define a 3-dimensional histogram over subject, predicate and object variables. Subject URIs are represented as strings[3] while predicate URIs are treated as categorical values, since there is always a small number of distinct predicates. Each bucket is composed of a 3-dimensional subject/predicate/object bounding box, a size indicating the number of triples contained in the bucket, and the number of distinct subjects, predicates and objects.

We experiment on a real query workload extracted from the logs of the user evaluation of the Semagrow Stack [21]. We separated the workload into a training set that is used to refine a histogram H over D and an evaluation set that is used to compare the statistics reported by the histogram against the actual dataset. Specifically, we measure the *average absolute estimation error* and the *root mean square error* of histogram H on the respective workload W:

$$\mathrm{err}_{H,D}^{ABS}(W) = \frac{1}{|W|} \sum_{q \in W} |\mathrm{est}_H(q) - \mathrm{act}_D(q)|$$

$$\mathrm{err}_{H,D}^{RMS}(W) = \frac{1}{|W|} \sqrt{\sum_{q \in W} (\mathrm{est}_H(q) - \mathrm{act}_D(q))^2}$$

where $\mathrm{est}_H(q)$ is the cardinality estimation for query q and $\mathrm{act}_D(q)$ is the actual number of tuples in D that satisfy q.

The expected behaviour of the algorithm is to improve estimates by adding buckets that punch holes and add sub-buckets in areas where there is a difference between the actual statistics and the histogram estimates. Considering how client applications access some 'areas' more heavily than others, the algorithm zooms into such critical regions to provide more accurate statistics. Naturally, the more interesting observations relate to the effect of merges as soon as the available space is exhausted, so we have allocated to STRHist unrealistically small memory (50 and 100 buckets).

---

[2] Please see http://agris.fao.org for more details on AGRIS. The AGRIS site mentions 7 million distinct publications, but this includes recent additions that are not in end-2013 data dump used for these experiments.

[3] We use the *canonical string representation* of URIs as defined in Sect. 2, IETF RFC 7320 (http://tools.ietf.org/html/rfc7320).

5.2   Results

The AGRIS workload queries follow the same template: Both subjects and predicate URIs are defined by the query, leaving the object dimension unrestricted. As it represents a real scenario, we may have duplicate queries in the workload. To generate the workload we randomly select a set of queries for refinement and another set for evaluation. Therefore, we create 24 batches of 55 training queries, totalling 1320 training queries, followed by a set of 100 evaluation queries used to compare the estimations against the actual size of the query results and the estimated ones. We experiment with different system configurations. Specifically, we

Table 1. Estimation error (RMS and absolute) versus training batch and merges (parent-child (PC) and sibling-sibling (SS) merges) using prefixes and similarity ranges. Configured for a maximum of 50 buckets.

Training batch	Similarity ranges					Prefix ranges				
	Error		Merges			Error		Merges		
	RMS	Abs	PC	SS	Total	RMS	Abs	PC	SS	Total
01	0.283	2.14	0	0	0	0.283	2.14	0	0	0
02	0.414	2.58	3	3	6	0.457	2.67	5	12	17
03	1.728	9.26	23	6	29	1.562	6.61	8	30	38
04	1.758	9.84	19	8	27	2.350	11.55	11	28	39
05	0.899	7.89	13	6	19	2.711	15.13	12	30	42
06	4.483	40.84	9	13	22	5.856	26.36	9	23	32
07	4.691	44.66	31	0	31	6.844	32.58	11	28	49
08	4.762	46.08	44	1	45	6.724	38.20	5	44	49
09	4.735	45.58	31	21	52	6.911	41.52	5	42	47
10	4.787	46.57	20	4	24	7.968	46.96	11	28	39
11	4.794	47.07	25	3	28	10.444	60.59	11	28	39
12	4.814	47.07	15	6	21	12.153	70.67	13	27	40
13	4.814	43.56	23	6	29	13.883	81.95	13	28	41
14	4.608	43.56	23	8	31	14.201	85.07	12	27	39
15	4.608	47.58	28	6	34	14.201	85.07	11	28	39
16	4.841	47.58	29	4	33	19.365	110.09	14	28	42
17	4.841	47.58	35	4	39	23.147	131.65	14	28	42
18	4.841	47.58	24	5	29	23.415	134.37	13	27	40
19	4.841	47.58	24	4	28	23.792	137.85	10	28	38
20	4.841	47.58	41	1	42	23.792	137.85	15	28	43
21	4.841	47.58	32	5	37	27.048	157.13	13	28	41
22	4.841	47.58	14	2	16	27.048	157.13	14	28	42
23	4.841	47.58	27	2	29	27.567	162.20	13	28	41
24	4.841	47.58	14	1	15	27.567	162.20	2	8	10

set a maximum of 100 and 50 buckets. Moreover, we evaluate both reported representations for string ranges (i.e. prefix ranges and similarity ranges). Tables 1 and 2 depict the average errors of the evaluation queryset and the number of merges performed during each training batch.

One can note that the similarity range approach produced more accurate estimations, especially when the maximum number of buckets is very limited causing more merges. Using this observation we can infer that the similarity range approach makes better merging decisions than the prefix range one. The reason that prefixes cannot create as good merged buckets as in the similarity

Table 2. Estimation error (RMS and absolute) versus training batch and merges (parent-child (PC) and sibling-sibling (SS) merges) using prefixes and similarity ranges. Configured for a maximum of 100 buckets.

Training batch	Similarity ranges					Prefix ranges				
	Error		Merges			Error		Merges		
	RMS	Abs	PC	SS	Total	RMS	Abs	PC	SS	Total
01	0.283	2.14	0	0	0	0.283	2.14	0	0	0
02	0.259	1.73	0	0	0	0.259	1.73	0	0	0
03	0.259	1.73	0	0	0	0.259	1.73	0	0	0
04	0.408	2.56	10	4	14	0.259	1.73	7	8	15
05	1.688	8.88	20	10	30	0.259	1.73	6	30	36
06	1.768	10.94	15	9	24	0.259	1.73	10	24	34
07	4.581	32.96	24	4	28	0.259	1.73	13	28	41
08	5.886	40.53	23	11	34	0.472	2.48	9	33	42
09	8.236	76.94	37	19	56	1.919	5.91	1	52	53
10	8.236	76.94	22	3	25	2.687	8.87	13	29	42
11	6.654	50.75	17	5	22	4.624	12.11	11	27	38
12	6.136	43.52	12	9	21	4.960	13.79	13	28	41
13	5.921	40.84	18	5	23	5.528	15.23	12	28	40
14	5.530	35.94	13	3	26	5.537	15.53	13	27	40
15	5.740	35.92	7	5	12	5.537	15.93	13	28	41
16	5.740	35.94	9	4	13	5.806	16.92	13	27	40
17	6.190	41.42	16	4	20	5.955	17.72	10	28	38
18	5.623	34.68	13	5	18	5.955	17.72	11	27	38
19	5.623	34.68	10	2	12	9.658	25.56	8	22	30
20	5.623	34.68	6	0	6	10.846	28.44	15	28	43
21	5.623	34.65	12	7	19	10.846	28.44	11	27	38
22	6.102	37.50	12	11	23	12.453	33.24	11	28	39
23	6.182	38.47	21	0	21	12.872	35.16	13	27	40
24	10.137	98.79	11	0	11	12.876	35.56	8	17	25

ranges is that (a) prefixes as a succinct description is more restrictive and (b) the AGRIS URIs have a hierarchical structure, but this structure is not that deep that it would make strict prefixes expressive. Notice that the total merges performed per batch are fewer in the similarity range case. This is due to the fact that more training queries are already accurately estimated and thus the histogram refinement algorithm discards them without drilling new holes. Moreover, this observation is also consistent even after considerable merges have been applied to the histogram, deducing that merged buckets are not introducing significant error to the estimations.

The histogram stabilizes after a certain number of training batches, as evidenced by the fact that the error remains constant. A significant difference can be seen in the type of merging preferred by the two approaches: the number of the parent-child merges is higher in similarity range approach, while the prefix range approach prefers the sibling merging. This demonstrates the bias towards parent-child merges encoded by the distance-based penalty in similarity merging.

6   Conclusions

In this article we have presented an algorithm for building and maintaining multi-dimensional histograms exploiting query feedback. Our algorithm is based on STHoles algorithm, but extends it to also handle URIs. One significant contributions of the article is that it establishes a framework that formalises histograms over arbitrary data types and identifies the specification of a language for specifying data ranges as a key element of histograms. Building upon this, we have identified the properties that any such language should have for histogram refinement algorithms to be applicable.

This led to the second major contribution, that of proposing the Jaro-Winkler similarity metric as an appropriate basis upon which to build a formalization of histograms over URI strings. This metric has the advantage of accommodating the hierarchical nature of URI strings by placing more importance on the beginning of the string, while being more flexible than strict prefix matching. This gives our system a great advantage over the state of the art, where ranges are only defined over numerical data and strings are treated as categorical data that can only be described by enumeration: by having the ability to succinctly describe ranges of related URI strings, finer (and thus more accurate) histograms can fit a given amount of memory.

As future work, we will experiment with a more sophisticated estimation of the size of a bucket based from the radius of its box. One idea would be to dynamically adapt a conversion ratio parameter to the observed query feedback, so as to better fit each given dimension in a dataset. This will improve multi-dimensional volume calculations, since it will lift the assumption that the breadth of a URI description and the size of the data that fits the description grow uniformly. An even more ambitious goal is to define the length of URI string ranges in a way that it can be combined with numerical range length, so that multi-dimensional and heterogeneous (strings and numbers) buckets can be assigned a meaningful volume.

Another strain of future research will experiment with finer representations of clusters of URIs than the radius around a single central URI. This would allow us to improve sibling merging, as our current approach is prone to over-generalizing and making the histogram sensitive to the query feedback it receives when it is first constructed.

With respect to the software development, we plan to develop a more scalable implementation of the algorithm which will be able to efficiently serve histograms from databases and not from in-memory Java objects. Although the unavoidable delay is not critical for the refinement phase, it can be unacceptable for the run-time usage of the histogram by query optimizers. To keep such delays manageable, a caching mechanism will need to be integrated in the implementation so that the most frequent accesses to the histogram are served from a memory cache.

Acknowledgements. The research leading to these results has received funding from the European Union's Seventh Framework Programme (FP7/2007-2013) under grant agreement No. 318497. For more details about the SemaGrow project please see http://www.semagrow.eu and about the Semagrow system please see http://semagrow. github.io.

References

1. Bruno, N., Chaudhuri, S.: Exploiting statistics on query expressions for optimization. In: Proceedings of the 2002 ACM International Conference on Management of Data (SIGMOD 2002), New York, NY, USA, pp. 263–274. ACM (2002)
2. Stillger, M., Lohman, G.M., Markl, V., Kandil, M.: LEO - DB2's LEarning optimizer. In: Proceedings of the 27th International Conference on Very Large Data Bases, VLDB 2001, San Francisco, CA, USA, pp. 19–28. Morgan Kaufmann Publishers Inc. (2001)
3. Aboulnaga, A., Chaudhuri, S.: Self-tuning histograms: building histograms without looking at data. In: Proceedings of the 1999 ACM International Conference on Management of Data (SIGMOD 1999), New York, NY, USA, pp. 181–192. ACM (1999)
4. Bruno, N., Chaudhuri, S., Gravano, L.: STHoles: a multidimensional workload-aware histogram. In: Proceedings of the 2001 ACM SIGMOD International Conference on Management of Data (SIGMOD 2001), pp. 211–222 (2001)
5. Srivastava, U., Haas, P.J., Markl, V., Kutsch, M., Tran, T.M.: ISOMER: consistent histogram construction using query feedback. In: Proceedings of the 22nd International Conference on Data Engineering (ICDE 2006), Washington, DC, USA. IEEE Computer Society (2006)
6. Roh, Y.J., Kim, J.H., Chung, Y.D., Son, J.H., Kim, M.H.: Hierarchically organized skew-tolerant histograms for geographic data objects. In: Proceedings of the 2010 ACM SIGMOD International Conference on Management of Data, SIGMOD 2010, New York, NY, USA, pp. 627–638. ACM (2010)
7. Kaushik, R., Suciu, D.: Consistent histograms in the presence of distinct value counts. Proc. VLDB Endowment **2**, 850–861 (2009)
8. Markl, V., Haas, P.J., Kutsch, M., Megiddo, N., Srivastava, U., Tran, T.M.: Consistent selectivity estimation via maximum entropy. VLDB J. **16**, 55–76 (2007)

9. Bruno, N., Chaudhuri, S., Weikum, G.: Database tuning using online algorithms. In: Liu, L., Özsu, M.T. (eds.) Encyclopedia of Database Systems, pp. 741–744. Springer, New York (2009)
10. Khachatryan, A., Müller, E., Stier, C., Böhm, K.: Sensitivity of self-tuning histograms: query order affecting accuracy and robustness. In: Ailamaki, A., Bowers, S. (eds.) SSDBM 2012. LNCS, vol. 7338, pp. 334–342. Springer, Heidelberg (2012). doi:10.1007/978-3-642-31235-9_22
11. Chaudhuri, S., Ganti, V., Gravano, L.: Selectivity estimation for string predicates: overcoming the underestimation problem. In: Proceedings of the 20th International Conference on Data Engineering (ICDE 2004), Washington, DC, USA. IEEE Computer Society (2004)
12. Lim, L., Wang, M., Vitter, J.S.: CXHist: an on-line classification-based histogram for XML string selectivity estimation. In: Proceedings of the 31st International Conference on Very Large Data Bases (VLDB 2005), Trondheim, Norway, 30 August – 2 September 2005, pp. 1187–1198 (2005)
13. Ding, L., Finin, T., Joshi, A., Pan, R., Cost, R.S., Peng, Y., Reddivari, P., Doshi, V., Sachs, J.: Swoogle: a search and metadata engine for the semantic web. In: Proceedings of the Thirteenth ACM International Conference on Information and Knowledge Management, CIKM 2004, New York, NY, USA, pp. 652–659. ACM (2004)
14. Auer, S., Demter, J., Martin, M., Lehmann, J.: LODStats – an extensible framework for high-performance dataset analytics. In: Teije, A., Völker, J., Handschuh, S., Stuckenschmidt, H., d'Acquin, M., Nikolov, A., Aussenac-Gilles, N., Hernandez, N. (eds.) EKAW 2012. LNCS (LNAI), vol. 7603, pp. 353–362. Springer, Heidelberg (2012). doi:10.1007/978-3-642-33876-2_31
15. Langegger, A., Wöss, W.: RDFStats - an extensible RDF statistics generator and library. In: 23rd International Workshop on Database and Expert Systems Applications, Los Alamitos, CA, USA, pp. 79–83. IEEE Computer Society (2009)
16. Harth, A., Hose, K., Karnstedt, M., Polleres, A., Sattler, K.U., Umbrich, J.: Data summaries for on-demand queries over linked data. In: Proceedings of the 19th International World Wide Web Conference (WWW 2010), Raleigh, NC, USA, 26–30 April 2010
17. Zoulis, N., Mavroudi, E., Lykoura, A., Charalambidis, A., Konstantopoulos, S.: Workload-aware self-tuning histograms of string data. In: Chen, Q., Hameurlain, A., Toumani, F., Wagner, R., Decker, H. (eds.) DEXA 2015. LNCS, vol. 9261, pp. 285–299. Springer, Heidelberg (2015). doi:10.1007/978-3-319-22849-5_20
18. Winkler, W.E.: String comparator metrics and enhanced decision rules in the Fellegi-Sunter model of record linkage. In: Proceedings of the Section on Survey Research Methods, Technical report, pp. 354–359. American Statistical Association (1990)
19. Charalambidis, A., Troumpoukis, A., Konstantopoulos, S.: SemaGrow: optimizing federated SPARQL queries. In: Proceedings of the 11th International Conference on Semantic Systems (SEMANTiCS 2015), Vienna, Austria, 15–18 September 2015
20. Charalambidis, A., Konstantopoulos, S., Karkaletsis, V.: Dataset descriptions for optimizing federated querying. In: Companion Proceedings of the 24th International World Wide Web Conference Companion Proceedings (WWW 2015), Poster Session, Florence, Italy, 18–22 May 2015
21. Celli, F., Keizer, J., Jaques, Y., Konstantopoulos, S., Vudragović, D.: Discovering, indexing and interlinking information resources. F1000Research **4** (2015). (Version 2; referees: 3 approved)

Author Index